Investigations in
EARTH SCIENCE

Lab Manual

Second Edition

Michael D. Bradley

Eastern Michigan University

Kendall Hunt
publishing company

Cover photos courtesy of Michael D. Bradley

www.kendallhunt.com
Send all inquiries to:
4050 Westmark Drive
Dubuque, IA 52004-1840

Copyright © 2009, 2011 by Michael D. Bradley

ISBN 978-1-4652-1402-7

Printed in the United States of America
10 9 8 7 6 5 4 3 2

Contents

Preface

This laboratory manual is designed to complement the ESSC 108 general education, introductory class in Earth Science for non-science majors offered at Eastern Michigan University. The material in this manual, in conjunction with your lecture textbook, provides the background information necessary to complete the lab exercises that are handed out at the beginning of each lab.

Each chapter provides an overview of the major concepts necessary to complete the laboratory. However, as with all laboratories, it is assumed that you have a thorough understanding of the theoretical concepts as presented in the ESSC 108 Earth Science lecture and textbook and are now ready for some practical applications of those concepts.

The vision for each of these exercises is to use the scientific method of inquiry to investigate an Earth Science subject. The exercises are designed to be investigatory in nature using common techniques used every day by Earth scientists. To assure that you are adequately prepared for each laboratory, each chapter has a PreLab associated with it. *You must complete the PreLab before you come to your lab.* The completed PreLab is your ticket to enter the lab.

This manual, being geared towards satisfying the general education laboratory-based science requirement at EMU, is not meant to be a comprehensive presentation of each topic but an application of a single (or few) concept(s) within the topic of interest.

Throughout the course, students will learn the procedures, practices, methodologies and assumptions that are fundamental to a scientific understanding of the Earth system. The overarching goal of the course is to create scientifically literate citizens that are both willing and able to participate responsibly in a global community by: (1) carefully applying the scientific method as a tool for problem solving, in general; (2) critically evaluating the scientific merit of anything that is presented as science (is it really science?), especially in the area of Earth System Science; and (3) thoroughly incorporating Earth System Science in important decisions and issues at the personal, local, national and global levels, such as where to live, where to store waste, and where and how to develop the surface of the planet.

Items to bring to each lab

To each lab you should bring this lab manual, your completed PreLab, the textbook from your lecture class, a calculator, a pencil, an eraser, and all previous graded labs since they will provide a reference on how to do some of the material that you may have forgotten.

Acknowledgements

The author wishes to acknowledge the comments and suggestions from the many laboratory instructors who have used this manual. I have made every effort to incorporate their teaching experiences into this revision.

I would also like to extend my gratitude to Todd Grote, Michael Kasenow, Tom Kovacs, and Serena Poli for their thoughtful reviews of this revision of the laboratory manual.

Introduction to Topographic Maps

INTRODUCTION

Maps are useful devises to show spatial relationships between one feature and another. Today there are many different kinds of maps used for many different purposes but for this exercise we are going to concentrate on **topographic maps**—maps that show the shape of the land. In addition to showing hills, valleys, rivers, mountains, etc., most topographic maps also show cultural features such as roads, towns, political boundaries, etc. Topographic maps show a **plan view** of Earth's surface (the view looking vertically down on Earth as from an airplane). Topographic maps include a coordinate system to determine direction and to locate geographic features and also include a variety of scales to measure horizontal distances. Lastly, topographic maps use contour lines to show changes in elevation. In this exercise you are going to work with each of these map elements.

There are myriad reasons you may find yourself using a topographic map. Topographic maps show the physiographic characteristics of the land (mountains, valleys, etc.) and the location of trails, water, shelter, and so on. so provide essential survival information to backpackers. If you are buying property, a topographic map will provide you with information of the lay of the land so you can determine how much material you may have to add or remove from an area to level out the property so you can build a house. You can also get an idea as to where water will collect by noting the slope of the land and the low lying areas. If you need to drill a water well you can use a topographic map to determine how deep it is to the water table by realizing that the surface of rivers and lakes are at the water table so you can just subtract the elevation of the closest body of water from the elevation of where you plan to drill your well to get an approximation of the depth to the water table beneath your property. GPS units provide locational information using the same system as displayed on topographic maps. Understanding what this information means will allow you to determine when you are getting accurate information and when you are not.

GENERAL MAP ELEMENTS

All topographic maps have a title (usually the name of the largest town or most prominent geographic feature in the area) located in the upper right and lower right corners of the map, reference to a location grid and cardinal points (north, east, etc.), a scale, and a date showing when the map was made. If the map has been updated using aerial photographs, the map will also have a photorevised date usually printed in purple ink just below the original date of publication. The location of each of these items is shown in figure 1.1.

The direction to the **geographic north** pole of Earth is indicated by a line with a star on the top or the letter "N". The direction to the **magnetic north** pole is indicated by a line with MN at the top. Geographic north and magnetic north do not coincide so the difference between the two, called **magnetic declination**, is also given. In figure 1.2 from the Hoyt Peak quadrangle, Utah magnetic north is 16° east of true (geographic) north. Topographic maps often use rectangular coordinate grids. Whereas meridians converge as you approach the North (or South) Pole, rectangular coordinates do not. Therefore there is a discrepancy between rectangular coordinate north (called **grid north** and labeled GN) and true north. In figure 1.2 grid north is 0° 07' west of geographic north. Information on the three "norths" is shown on a declination diagram along the base of the map.

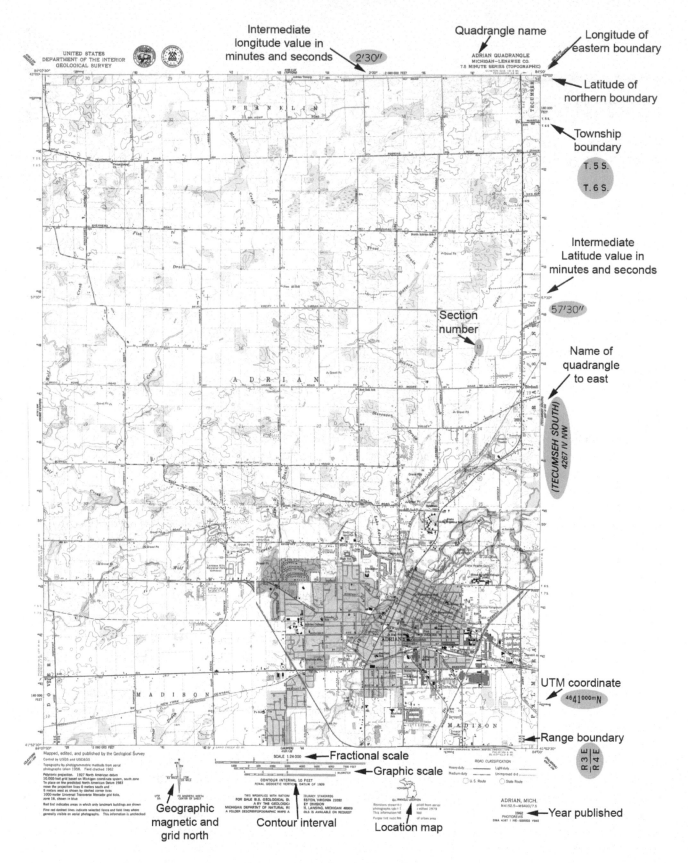

Intermediate longitude value in minutes and seconds — 2'30"

Quadrangle name

Longitude of eastern boundary

Latitude of northern boundary

Township boundary

T. 5 S.
T. 6 S.

Intermediate Latitude value in minutes and seconds

57'30"

Section number

Name of quadrangle to east

(TECUMSEH SOUTH) 4267 IV NW

UTM coordinate

4641 000m N

Range boundary

R 3 E / R 4 E

Fractional scale

Graphic scale

Geographic magnetic and grid north

Contour interval

Location map

Year published

FIGURE 1.1 ▶ *General features of topographic maps.*

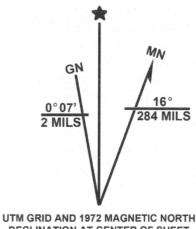

UTM GRID AND 1972 MAGNETIC NORTH
DECLINATION AT CENTER OF SHEET

FIGURE 1.2 ► *Geographic (true), magnetic, and grid north. In this case, the magnetic declination is 16° east of true north.*

The location of Earth's magnetic north pole slowly drifts from year to year so the magnetic declination shown on the map is accurate only for the year indicated on the map. You can get up-to-date information on the internet by searching on "magnetic declination".

COORDINATE AND LAND PARTITIONING SYSTEMS

All modern maps have a coordinate system. In the United States topographic maps typically have three separate coordinate systems, the worldwide system of latitude and longitude, the United States Public Land Survey (USPLS) used in the U.S. outside of the original 13 colonies, and the Universal Transverse Mercator (UTM) system used by the military. For this exercise you will use latitude and longitude and the USPLS systems.

LATITUDE AND LONGITUDE

A two-axis coordinate system is used for determining locations on the surface of Earth; one axis runs east-west and the other axis north-south. East-west lines are called **parallels** because they are all parallel to each other. Parallels are used to measure north and south (figure 1.3a). North-south lines that pass through Earth's poles are called **meridians**. Meridians are farthest apart at the equator and converge at the north and south poles. Meridians are used to measure east and west (figure 1.3b). The

pattern of meridians and parallels on the earth is called the **graticule** (figure 1.3c).

It is common practice to measure distances along the perimeter of a circle or sphere in degrees of arc with 360° (degrees) making a complete circle. The equator is half way between the North and South Poles so it is a logical origin to measure north and south degrees. Degrees north or south of the equator are

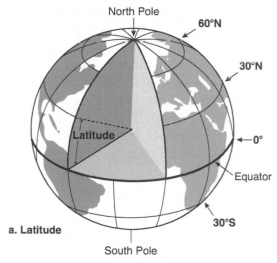

a. Latitude

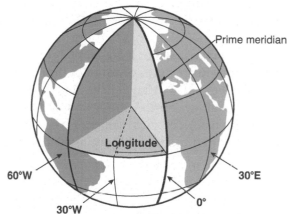

b. Longitude

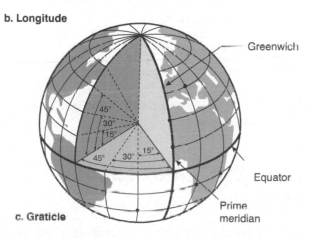

c. Graticle

FIGURE 1.3 ► *Latitude and longitude. See text for discussion.*

Chapter 1 *Introduction to Topographic Maps*   3

called degrees of **latitude**. Since the arc from the equator to a pole is one-quarter of a circle, latitudes can only vary from 0° at the equator to 90° at the poles. Degrees north of the equator are labeled "north". Degrees south of the equator are labeled "south".

The meridian that passes through the Royal Observatory in Greenwich, England arbitrarily serves as the starting point for measuring east and west degrees so is called the **prime meridian**. Degrees east or west of the prime meridian are called degrees of **longitude**. Degrees east of the prime meridian are labeled "east". Degrees west of the prime meridian are labeled "west". The half way point around the world from the prime meridian is called the **International Date Line**. Since this meridian is in the middle of the Pacific Ocean where few people live it is used to change the date from one day to the next.

By convention, degrees of arc are subdivided into 60' (minutes) of arc and minutes of arc are subdivided into 60" (seconds) of arc. The precise latitude and longitude is given for the four corners of a topographic map. In figure 1.4 the northeastern corner of the Adrian quadrangle is Latitude 42° 00' 00" N. and Longitude 84° 00' 00" W. Note that north and west are not stated on the map but it is assumed you know whether or not you are north or south of the equator and east or west of England. The conterminous United States is always north latitude and west longitude. In addition, intermediate values of latitude and longitude are given along the margins of the map (figure 1.1).

ADRIAN QUADRANGLE
MICHIGAN—LENAWEE CO.
7.5 MINUTE SERIES (TOPOGRAPHIC)

FIGURE 1.4 ▶ *The northeastern corner of the Adrian quadrangle, Michigan.*

U.S. PUBLIC LAND SURVEY SYSTEM

In 1785, Congress passed a land ordinance Act that provided the initial framework for a rectangular coordinate system for use in disposing of lands in the "Western Territory". This system has come to be called the United States Public Land Survey (USPLS). It started in Ohio and extends throughout most of the western U. S. and part of the south encompassing 30 states. In any coordinate system you must first establish a starting point so you can develop a reference grid. In the case of the USPLS there are 35 starting points across the mid-west and western parts of the United States (figure 1.5).

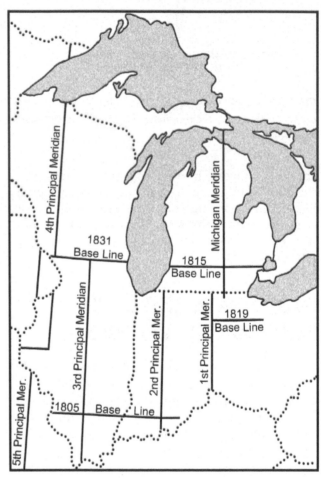

FIGURE 1.5 ▶ *USPLS reference grid for the upper Midwest U. S.*

Taking a cue from the longitude and latitude system, the north-south reference line in the USPLS system is called the **Principal Meridian** (figure 1.6). The east-west reference line is called the **Baseline**. A grid system is set up at six mile intervals both east and west of the principal meridian and north and south of the baseline. The rows are referred to as **Townships**. The first row north of the baseline is called Township 1 North (T. 1N.). The second row north of the baseline is called Township 2 North. The first row south of the baseline is called Township 1 South, and so on. The

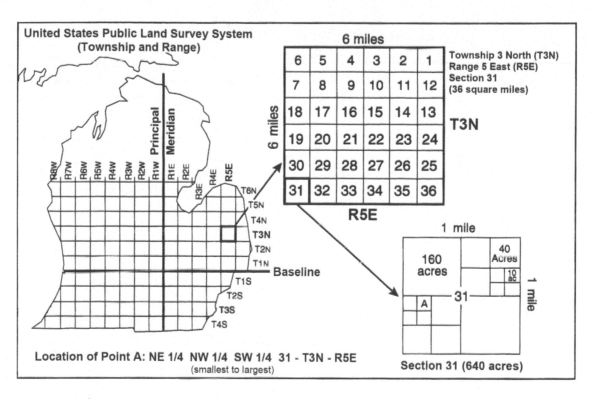

FIGURE 1.6 ▶ *The United Sates Public Land Survey System USPLS system.*

columns are referred to as **Ranges**. The first column east of the Principal Meridian is called Range 1 East (R. 1E.). The second column east of the Principal Meridian is called Range 2 East. The first column west of the Principal Meridian is called Range 1 West, and so on.

The grid sets up 36 square mile areas that are 6 miles on a side. Each Township is subdivided into 36 blocks of one square mile each (640 acres). These blocks are called **Sections**. Sections are commonly numbered 1-36 with 1 being in the northeast corner. Section 2 is immediately west of Section 1. When the west (or east) edge of the Township is reached, the numbering drops to the row immediately to the south. The numbering goes back and forth until Section 36 lies in the southeastern corner of the Township. See figure 1.8 for details.

The labels for Townships, Ranges and Sections are commonly in red letters. Townships are labeled along the north-south neatlines of the map, Ranges are labeled along the east-west neatlines of the map, and sections are labeled in the middle of the section. **Neatlines** are the lines that form the border of the map. Township and Range boundaries are generally drawn as bold red lines whereas section boundaries are thin red lines. If the township or range boundary is also a road or a county or state boundary, a bold

red line might not be present on the map but the boundary will still be labeled along the neatlines.

Sections, as recorded on plats, are subdivided in **Aliquot Parts** (figure 1.6). Aliquot Parts result from the division of sections into quarters (160 acres). Each of those quarters are subsequently subdivided into quarters (40 acres) and each of those quarters are subsequently subdivided into quarters (10 acres). In the example shown in figure 1.6 the "A" is in the NE¼ of the NW¼ of the SW¼ of Section 31, T3N, R5E. Notice the location is written from smallest to largest.

The Aliquot Parts system is not an accurate system for precise locations but it is how land in the U.S. was originally parceled by the government. When you look at a topographic map, note that roads tend to be along section boundaries and the quarter section boundaries. This makes sense if you think about property ownership. The roads tend to be along property boundaries which were originally formed using Aliquot Parts.

Defining a precise location such as the northeast corner of your property is not very practical using Aliquot Parts, so footage from a recognizable location is often used. The location of the house in figure 1.7 is 2050 feet from the west line (FWL) and 2550 feet from the south line (FSL) of Section 7, the "lines" being the section boundary lines. Note that using footages can give you can exact point.

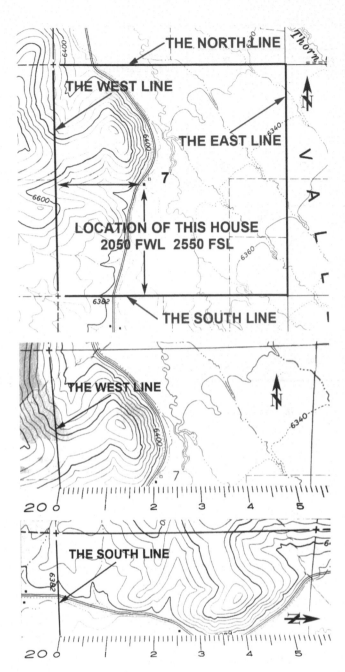

FIGURE 1.7 ▶ *Determining exact footages from the north, east, south, or west boundaries of a section. The scale of this map is 1:24,000 so one inch equals 2000 feet. The rulers shown are "20 scale" rulers where an inch is divided into 20 equal parts so each division on the ruler represents 100 feet (2000 feet/20 = 100 feet). The location of the house is 2050 feet from the west line (FWL) and 2550 feet from the south line (FSL) of Section 7, the "lines" being the section boundary lines. Note that the bottom map has been rotated.*

SCALE

The **scale** of a map is the ratio between map distance and earth distance. There are three ways that scale is commonly stated: word statement, representative fraction, and graphic scales.

A **word statement** is a written description of map distance, such as "one inch on the map equals one mile in the real world" or "one inch equals 2000 feet", and so on. Notice that in word statements mixed units (inches and feet, inches and miles, etc.) are common.

The ratio between the map distance and the ground distance between equivalent points is called the **Representative Fraction (RF)**. RF is stated as a true ratio, 1:24,000 or as a fraction, 1/24,000 both mean the same thing. In figure 1.7, one unit on the map equals 24,000 units on the earth. So one inch on the map equals 24,000 inches on the earth or one centimeter on the map equals 24,000 centimeters on the earth. If you divide a pie into four parts, those pieces are larger than if you divided the same pie into eight parts. Thus, for a fractional scale, a smaller number in the denominator is a larger scale map (more detailed) than a map with a larger number in the denominator; so a 1:24,000 scale map is more detailed than a 1:250,000 scale map.

The representative fraction scale is only accurate on the original map. If you enlarge or reduce the map, the representative fraction is no longer accurate. This sets up the basis for the graphic scale.

A line or bar drawn on a map that directly relates map scale to earth scale is called a **graphic scale**. For example, if the map scale was one inch equals 2000 feet, then one inch on the graphic scale would be labeled 2000 feet. Often several bars are drawn with scales in miles, feet, and kilometers. One advantage to a graphic scale is the line length will enlarge or reduce if you enlarge or reduce the map so the graphic scale will always be correct. Copy machines produce a slightly distorted image so for precise measurements you should always use originals.

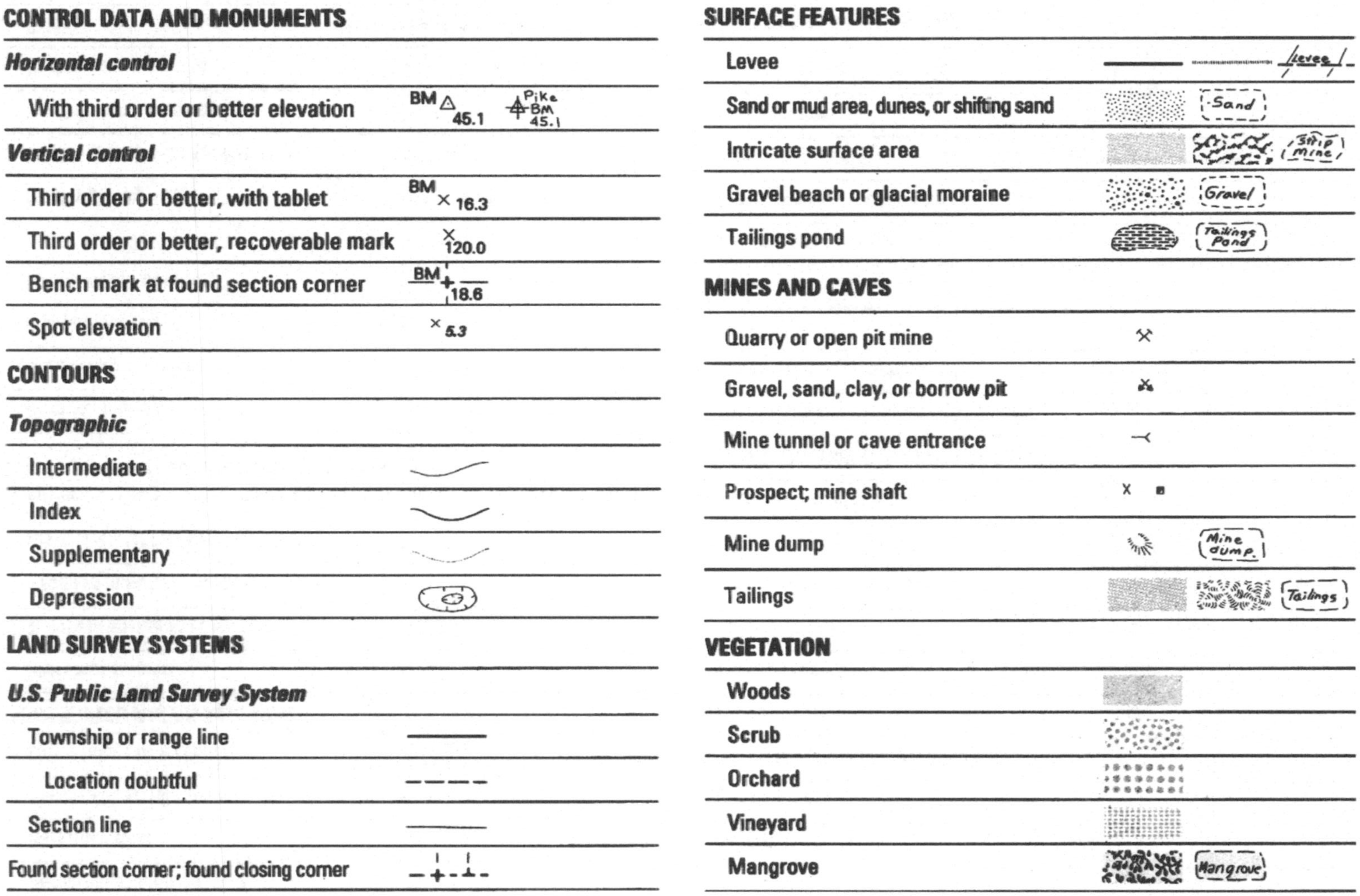

CONTROL DATA AND MONUMENTS

Horizontal control

With third order or better elevation   BM △ 45.1   Pike △ BM 45.1

Vertical control

Third order or better, with tablet   BM × 16.3

Third order or better, recoverable mark   × 120.0

Bench mark at found section corner   BM + 18.6

Spot elevation   × 5.3

CONTOURS

Topographic

Intermediate

Index

Supplementary

Depression

LAND SURVEY SYSTEMS

U.S. Public Land Survey System

Township or range line

Location doubtful

Section line

Found section corner; found closing corner

SURFACE FEATURES

Levee

Sand or mud area, dunes, or shifting sand   (Sand)

Intricate surface area   (Strip mine)

Gravel beach or glacial moraine   (Gravel)

Tailings pond   (Tailings Pond)

MINES AND CAVES

Quarry or open pit mine

Gravel, sand, clay, or borrow pit

Mine tunnel or cave entrance

Prospect; mine shaft

Mine dump   (Mine dump)

Tailings   (Tailings)

VEGETATION

Woods

Scrub

Orchard

Vineyard

Mangrove   (Mangrove)

FIGURE 1.8a ▲ *Explanation of topographic map symbols.*

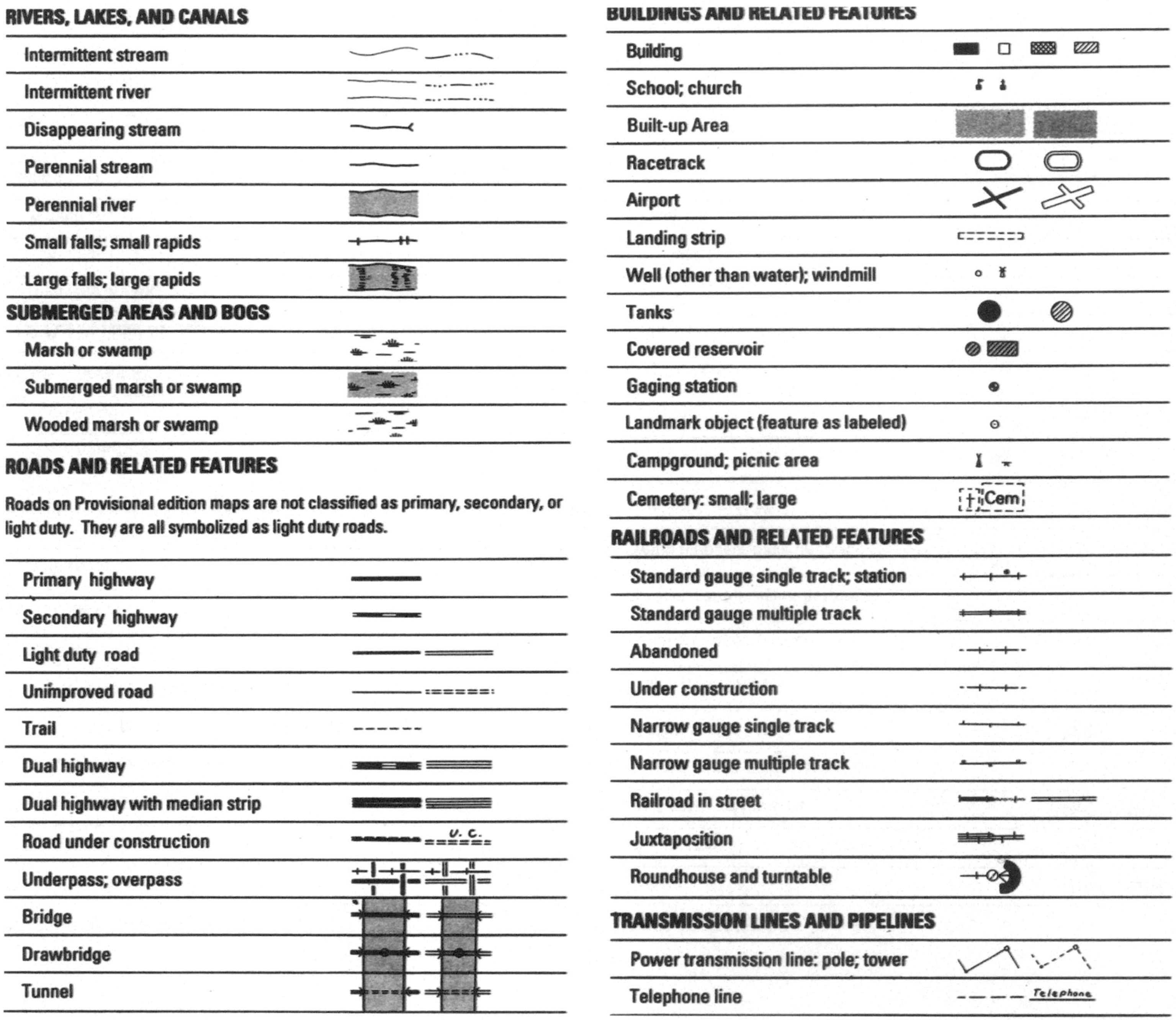

RIVERS, LAKES, AND CANALS

Intermittent stream

Intermittent river

Disappearing stream

Perennial stream

Perennial river

Small falls; small rapids

Large falls; large rapids

SUBMERGED AREAS AND BOGS

Marsh or swamp

Submerged marsh or swamp

Wooded marsh or swamp

ROADS AND RELATED FEATURES

Roads on Provisional edition maps are not classified as primary, secondary, or light duty. They are all symbolized as light duty roads.

Primary highway

Secondary highway

Light duty road

Unimproved road

Trail

Dual highway

Dual highway with median strip

Road under construction

Underpass; overpass

Bridge

Drawbridge

Tunnel

BUILDINGS AND RELATED FEATURES

Building

School; church

Built-up Area

Racetrack

Airport

Landing strip

Well (other than water); windmill

Tanks

Covered reservoir

Gaging station

Landmark object (feature as labeled)

Campground; picnic area

Cemetery: small; large

RAILROADS AND RELATED FEATURES

Standard gauge single track; station

Standard gauge multiple track

Abandoned

Under construction

Narrow gauge single track

Narrow gauge multiple track

Railroad in street

Juxtaposition

Roundhouse and turntable

TRANSMISSION LINES AND PIPELINES

Power transmission line: pole; tower

Telephone line

FIGURE 1.8b ▲ *Explanation of topographic map symbols.*

PreLab

WORKSHEET

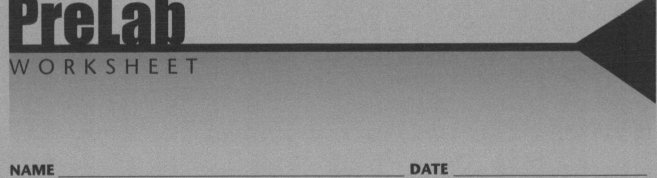

NAME _____ **DATE** _____

In what two corners on the map (upper left, upper right, lower left, lower right) is the name of the quadrangle located?

What is the reference by which longitude is measured?

Is Michigan's longitude north, south, east, or west?

What is the reference by which latitude is measured?

Is Michigan's latitude north, south, east, or west?

The topographic map below is at a scale of 1:24,000. The ruler shown on the map is a "20 scale" ruler meaning that one inch is divided into 20 equal parts. What is the distance (in feet) between building "A" and building "B"?

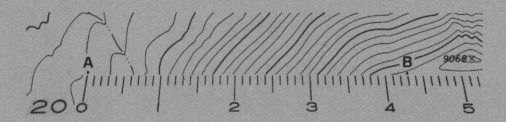

Using the explanation of "Topographic Map Symbols" what does the symbol below represent? The symbol is shown actual size at the bottom of the figure and enlarged for easier viewing at the top of the figure.

Chapter 2

Contouring, Slope, and Topographic Profiles

CONTOURING

Contours are lines connecting points of equal value on a single surface, such as elevation on the surface of the earth. Contouring is a means to represent topography (the third dimension) on a two-dimensional map. By convention, mean sea level is used as the datum (starting point) to determine elevation.

The concept of using contours to represent topography on Earth's surface is relatively straightforward and easy to illustrate using the changing shoreline of an island during a time of changing water level (figure 2.1).

The trace of the shoreline of an island marks points of equal elevation. So the shoreline of an island in the ocean is a contour line representing an elevation of zero (sea level). If sea level were to rise by 10 feet, the water would travel up the shore to an elevation 10 feet higher than the original shoreline. The new shoreline is a contour line representing an elevation 10 feet higher than the original shoreline. If sea level rose an additional 10 feet, the new shore line would represent an elevation 20 feet higher than the original shoreline, and so on.

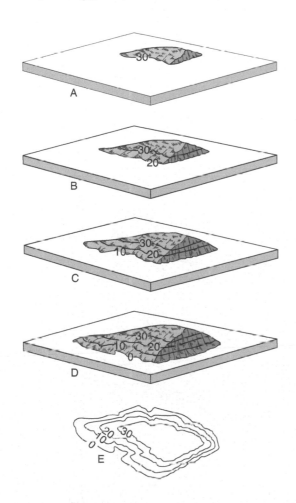

FIGURE 2.1 ▶ *Illustration of the concept of contour lines. See text for discussion.*

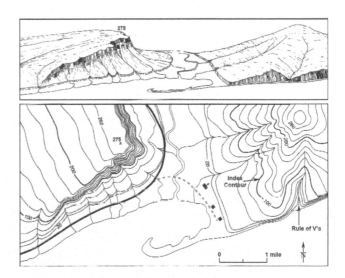

FIGURE 2.2 ▶ *Perspective and topographic map views of an area.*

Figure 2.2 shows a perspective view of an area and a topographic contour map depicting the same area. Note that since contours represent elevation changes, contour lines are closer together on steep slopes and farther apart on gentle slopes.

The change in elevation represented by adjacent contour lines is referred to as the **contour interval**. A single contour interval is used for the entire map. In figure 2.2, the contour interval is 20 feet. The contour interval varies from map to map and is usually chosen to best show the topography. In the United States 5, 10, 20, 40, or 80 feet are common contour intervals used on topographic maps.

Every fifth contour is drawn as a heavy line to facilitate reading. This contour is referred to as an **index contour**. Index contours are generally labeled with the elevation of the contour line (figure 2.2). The other contours (drawn with a thinner line) are called intermediate contours.

It is common to construct contour maps from point data. Figure 2.3 illustrates how to construct a contour map from point data. Contouring point data is accomplished by following a few simple rules.

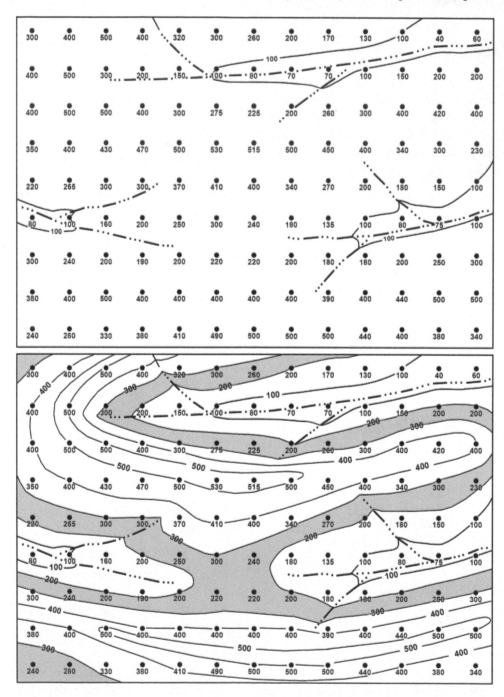

FIGURE 2.3 ▶ *Point data (top) and contoured point data (bottom). See text for discussion.*

Since contours are lines of equal elevation they can never cross each other or a single contour can never split into two parts. If two contours crossed each other, it would mean a single location had two different elevations. A single contour splitting into two or more contours would, likewise, represent an impossible situation. Occasionally, on a topographic map contours may appear to merge to express a vertical cliff but in reality they are stacked one on top of the other and only appear to touch.

All data between two contours must have values that lie between the values of the two contours. To illustrate this point the area between the 200-foot contour and the 300-foot contour in figure 2.3 has been shaded-in. Note that all point data within the shaded areas have values between 200 and 300.

Contouring involves trial and error. Note that adjacent contours tend to subparallel each other, so as you begin drawing contours you will find that altering the position of one line may require that several adjacent lines be adjusted. Contouring involves lots of erasing as you continuously reinterpret the data, so make sure you have a big eraser.

The bottom of Figure 2.3 shows an interpretation of the data. Note the word "interpretation"; contours are an interpretation of the data. The data may support multiple interpretations so your final map will differ in detail from other interpretations.

Contours are sequential and move in a linear progression (do not skip contours). A change in contour direction requires a repeat of the last contour. In figure 2.4a, if you pass through the 250 foot contour to get to the top of the ridge, you must pass through the 250 foot contour again when you come down the other side of the ridge. A consequence of this is that hills are represented by a series of closed contour lines. Depressions look similar except hachures (short lines) are placed on the downhill side of the contours (figure 2.4b). In fact, all contours are closed loops though closure often happens out of the map area.

In figure 2.2 note that all contour lines trend up valleys, cross the stream (if present), and extend down the valley on the opposite side. Thus, the contour lines form a "V" pattern with the apex of the "V" pointing up stream (called **Rule of V's**). This is true for any ravine whether or not a stream exists in the ravine.

Contouring is your interpretation of the data. In contouring, the guiding principle should be a topo-

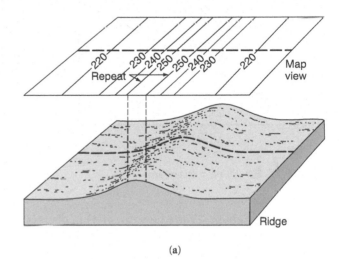

FIGURE 2.4 ▶ *See text for discussion.*

graphic one not a mathematical one. The final map should be a sensible picture of a dynamic regime caught in an instant of time. This "picture" of a local area should be consistent with the known regional topography. Each map is unique and provides an intimate look at the interpreter. Does your contour map show well-formed, coherent, and incisive ideas or does the interpretation show little thought or sense of direction? At this stage you'll be doing good to just honor the data but keep in mind that the quality of your work says a great deal about who you are.

SLOPE

A slope is a change in elevation over some distance. Slopes are generally expressed in three alternative

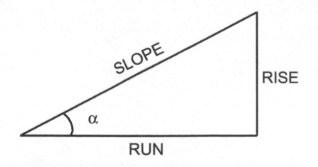

FIGURE 2.5 ▶ *Slope = Rise/Run.*

ways: 1) as a gradient, 2) as a percent, or 3) in degrees. The same slope can be expressed in any of these three forms. The particular form chosen is simply a matter of convenience in relation to the task at hand.

The **gradient** of a slope is the fraction that represents the ratio between the vertical distance, known as the rise, and the horizontal distance, known as the run. The rise may also be referred to as the **relief**—the difference in elevation between any two points.

Both rise and run must be stated in the same units, whether the units used are feet, meters, inches, or yards. To compute this ratio, a starting point and ending point are located on the map. The elevation of each point is determined using the contours and the horizontal distance between them is determined using the map scale. The slope is then written as the ratio between the (vertical) rise and the (horizontal) run, with the ratio reduced to show the horizontal distance per unit of vertical distance.

For example, calculate the gradient between point X and point Y in figure 2.6. The rise (relief) between X and Y is 30 feet (from the contour lines, the elevation of point Y, 1000 feet minus the elevation of point X, 970 feet). If the horizontal distance between X and Y is 1200 feet (use the map scale to determine this), the slope is 30/1200, which reduces to 1/40, or forty units of run per unit of rise. Note that, the slope is expressed as a ratio, so the slope designation is unit free.

Slope can be expressed as a **percent** by simply determining the rise per 100 units of run (divide the rise by the run and multiply the resulting decimal fraction by 100). In the previous example, the rise was 30 feet, and the run was 1200 feet. This is equivalent to a 2.5 percent slope (30 feet /1200 feet × 100% = 0.025 × 100% = 2.5%).

The third means of expressing slope is as an angle, in **degrees** from the horizontal. The angle of slope may be measured directly with a protractor from a topographic profile drawn with no vertical exaggeration or calculated using trigonometry.

TOPOGRAPHIC PROFILES

A **profile view** is the view you see as you look out horizontally across the landscape. A profile view can be constructed from a topographic map. First, select a line along which you would like to draw the topographic profile. In figure 2.6 the line **A – B** is used to construct a topographic profile. Elevation data from each location where a contour intersects the profile line (A – B) is transferred to the topographic profile. Once all the elevation data have been transferred to the topographic profile connect the data points to form the topographic profile.

If the vertical scale used for the topographic profile is not the same as the horizontal scale used on the topographic map, the topographic profile will be vertically exaggerated. It is common to use vertical exaggeration on topographic profiles to make the topography easier to see.

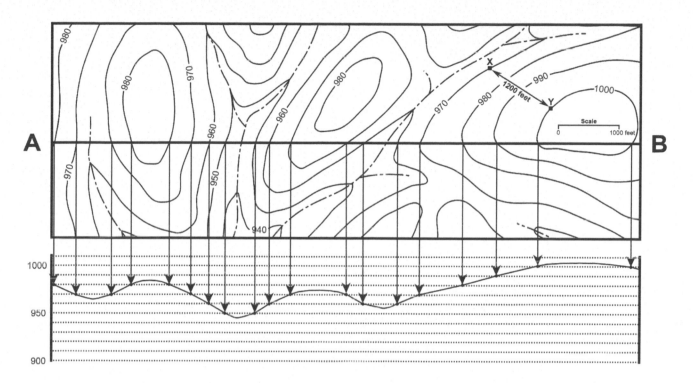

FIGURE 2.6 ▶ *Constructing a topographic profile. See text for discussion.*

NAME _____ DATE _____

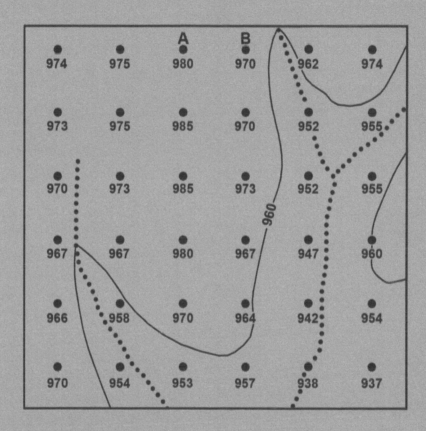

Figure 2.7 contains elevation data (in feet) at several locations (point data). The dotted lines are streams. The 960-foot contour has been drawn for you. Draw the 940-, 950-, 970-, and 980-foot contours. Remember the Rule of V's when contours cross the streams. All streams flow from higher elevations to lower elevations.

What is the contour interval on figure 2.7?

What is the relief between points A and B on figure 2.7?

If the horizontal distance between point A and point B is 200 feet, what is the gradient of the slope between A and B? Give your answer as a fraction.

Chapter 3

Mineral Identification

INTRODUCTION

In this exercise you are going to determine the physical properties of several different minerals. Then you will use these physical properties to identify each mineral by comparing your data with the physical properties characteristic of each mineral as provided on a Mineral Identification Key. Bear in mind that some of these tests are very precise but others are more subjective. For example, what you may call honey colored, someone else may describe as yellow. This type of subjectivity is built into the identification process so will not be a problem. The important thing to keep in mind is that generally no single property is diagnostic so you must perform several tests to positively identify a given mineral.

BACKGROUND

Minerals are the fundamental building blocks of the solid Earth. Therefore, to understand Earth and Earth processes, you need to first understand minerals. A **mineral** is a naturally occurring inorganic solid with a definite internal structure and a definite chemical composition that varies only within strict limits. Internal structure and chemical composition determine a mineral's physical properties. All specimens of a given mineral have the same physical properties regardless of when, where, or how formed, so physical properties can be used to identify a given mineral. In this exercise, you will learn several physical properties commonly used to identify minerals.

PHYSICAL PROPERTIES OF MINERALS

The physical properties of minerals that you can use to identify a mineral are summarized in the Mineral Identification Key.

Crystal Habit

If a mineral is allowed to grow freely without being restricted by adjacent crystals, it will assume a specific crystal shape inherent to its internal structure. The crystal shape is an expression of the atomic arrangement. This does not mean that every crystal of a given mineral will be the same size and shape since very few crystals grow in an unrestricted environment. What it means is that each mineral has a characteristic crystal form; though size and shape of the mineral may vary, the angles between crystal faces are constant. This property is referred to as Nicolaus **Steno's Law of Constancy of Interfacial Angles**.

If you had a perfect crystal that was large enough to measure the angles between the crystal faces, you could use this information to help identify the mineral. However, from a practical standpoint, most crystals do not grow in an unrestricted environment so are not perfect; and few minerals are large enough to accurately measure interfacial angles.

Color

The color of a mineral is an important property but it rarely is a definitive property. The reason is that many minerals contain impurities that also can affect the color. For instance, the common mineral quartz will be clear and glass-like if it is 99.998% pure SiO_2, pink (rose quartz) if it contains just 0.003% titanium as an impurity (that is just three Ti atoms per 100,000 SiO_2 molecules), and purple (amethyst) if it contains 0.020% iron as an impurity. Impurities are so common in some minerals that the color of the mineral is not a reliable diagnostic property in which to identify the mineral.

Streak

Geologists refer to the color of a mineral in its powdered form as the mineral's **streak**. If you grind up a mineral into a fine powder, sometimes the color of the mineral powder is different than the color of the whole mineral. It is impractical to take a large hammer and pound a mineral into dust because striking a mineral with a hammer would most likely send chips flying in all directions. The safe way to grind a mineral into a fine powder is to rub the mineral across a hard porcelain plate. Pieces of the mineral will rip off and be deposited as a streak of powder on the porcelain plate. Streak can be a reliable diagnostic property for some mineral identification. If the mineral is harder than the streak plate, the mineral will gouge the streak plate so any residue you see is powder from the streak plate not the mineral. After you run a streak test, check the streak plate to see if it was scratched. If the plate is scratched, the streak test cannot be used for this mineral. It is easier to see a light colored streak if you use the black streak plate and a dark colored streak if you use the white streak plate.

Luster

Luster is the appearance of light reflected from the surface of the mineral. Luster is described by comparing the mineral's appearance to common objects. For example, if the surface of the mineral looks like a bright, shiny metal, the mineral is said to have a metallic luster. In porcelainic luster, the surface of the mineral looks like a piece of unglazed porcelain, like the streak plate. In vitreous (glassy) luster, the surface of the mineral looks like a piece of glass. In resinous luster, the surface of the mineral looks like resin (like a piece of amber [fossilized tree sap]). In pearly luster, the surface of the mineral looks like a pearl. In dull luster, the surface of the mineral is dull or earthy in appearance.

These terms are subjective so what you would call vitreous luster, someone else may call porcelainic luster so be aware of this when using the Mineral Identification Key.

Hardness

Hardness is a measure of the resistance of a mineral to abrasion or scratching. Geologists use a standardized scale developed by Mohs to rank the hardness of a mineral. Mohs' hardness scale (figure 3.1) is a relative scale in which we compare the hardness of a given mineral to ten common minerals that range in hardness from very soft to very hard.

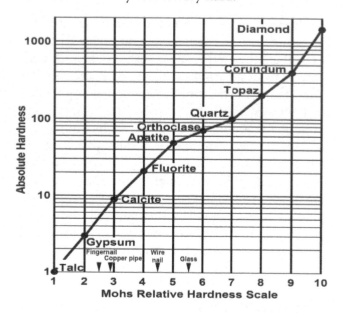

FIGURE 3.1 ▶ *Mohs' hardness scale showing the ten Mohs' minerals and also the hardness of a few common objects.*

To determine a mineral's hardness, you rub the mineral you are trying to identify against one of the minerals on Mohs' Hardness Scale. If the mineral you are trying to identify scratches the Mohs' mineral, then you know your unknown mineral is harder than the Mohs' mineral you scratched. You repeat the test using several different Mohs' minerals until you can determine the hardness of the mineral you are trying to identify. For example, if your unknown mineral scratched orthoclase but did not scratch quartz then you would know that your mineral is harder than orthoclase (6) but softer than quartz (7), therefore you can conclude that your unknown mineral has a hardness between 6 and 7 on Mohs' hardness scale.

It is obvious that if you were doing a lot of mineral identification you would use up a large box of

Mohs' minerals. Therefore, we substitute common items for Mohs' minerals. Glass has a hardness of 5.5, so the first hardness test is to see if the mineral will scratch a glass plate since this will tell you if the mineral is 1–5 or 6–10 on Mohs' hardness scale. Other common objects used in mineral hardness tests, are a human fingernail, which has a Mohs' hardness of 2.5, and a copper pipe, which has a Mohs' hardness of 2.9.

Cleavage and Fracture

Cleavage is the tendency of a mineral to split along planes of bonding that exist between atoms in the crystalline structure. Cleavage planes reflect the internal structure of a mineral and are characteristic for a given mineral. Cleavage is described as **perfect** if it results in flat planes. If a cleavage is not perfect, the plane will tend to be a rougher surface.

FIGURE 3.2 ▶ *Halite (rock salt) showing three directions of cleavage.*

A mineral can have more than one direction of cleavage. The mineral halite (figure 3.2) has three directions of cleavage and each of these planes of bonds are 90º to each other so halite tends to break into rectangular shapes.

If there are no planes of bonds, a mineral will not cleave but fracture along a non-planar surface, in some cases, just like glass breaks (conchoidal fracture). Note that in conchoidal fracture (figure 3.3) the fracturing occurs along smooth curving surfaces.

Distinguishing between cleavage and crystal habit (shape) is a destructive process since you have to break the mineral. If you break a mineral and the pieces retain the flat surfaces in the same orientations

FIGURE 3.3 ▶ *Conchoidal fracture.*

as the original mineral, the shape of the mineral was due to its cleavage. The left side of figure 3.4a is halite showing three cleavages at 90° to each other. The center frame shows the mineral being broken by a mallet. In the right frame, note that the broken pieces still exhibit three cleavages at 90° to each other.

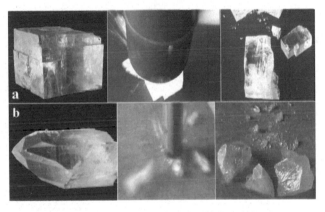

FIGURE 3.4 ▶ *Cleavage versus crystal habit. See text for discussion.*

If you break a mineral and the pieces have a different shape than the original mineral, the shape of the mineral was due to its crystal habit. The left side of figure 3.4b is a euhedral (perfect) quart crystal, exhibiting several flat crystal faces. The center frame shows the quartz crystal being broken by a steel chisel. In the right frame, note that the broken pieces exhibit conchoidal fracture and do not have the same flat surfaces of the original quartz crystal; so you can conclude that the flat surfaces were not cleavage planes.

You will not be allowed to break the mineral specimens, so how will you determine if the shape of the mineral specimen in the lab kits is due to its cleavage or crystal habit? Euhedral crystals of quartz are the only minerals in the lab in which the mineral's flat surfaces are due to its crystal habit.

Density

Density is the mass of a mineral divided by the volume of the mineral. To determine a mineral's density, precisely weigh the mineral using a triple beam balance (figure 3.4). The triple beam balance used for this lab measures weight in grams. The center scale ranges from zero to 500g in 100 gram increments. The rear scale ranges from zero to 100 grams in 10 gram increments. The front scale ranges from zero to 10 grams in 0.1 gram increments. The triple beam balance must always be perfectly balanced to obtain an accurate reading. When the triple beam balance is perfectly balanced, the white line on the beam will line up with the adjacent white line at "0".

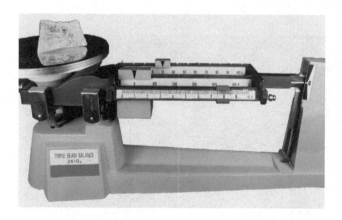

FIGURE 3.5 ▶ *Triple beam balance.*

To measure the precise volume of an irregularly shaped mineral, fill a graduated cylinder about half full of water. Note the precise volume of water in the cylinder. Drop the mineral into the water making sure the mineral is completely immersed. The water level will rise by the exact volume of the mineral. Note that the water surface in the graduate cylinder is curved. This curve is referred to as a **meniscus**. Read the bottom of the meniscus for an accurate reading (figure 3.5).

One milliliter of water occupies one cubic centimeter of volume so it is straightforward to determine a mineral's density in grams/cm³. The density of a mineral is its weight in grams divided by its volume in cubic centimeters.

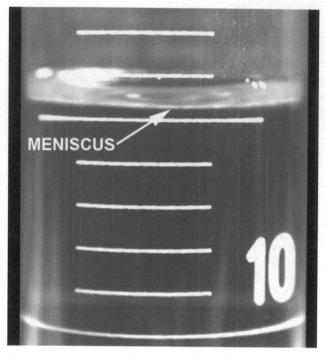

FIGURE 3.6 ▶ *Read the bottom of the meniscus when determining volume.*

Other Properties

Some minerals dissolve in dilute hydrochloric acid (HCl). Other tests that may be of value are taste (halite is salty), feel (graphite feels greasy), and magnetism (some minerals are magnetic).

Mineral identification key

Minerals with a Metallic Luster

Hardness	Density	Color	Streak	Cleavage/Fracture	Other Properties/Comments	Mineral
1.5–2	2.09–2.23 avg 2.16	iron black, dark gray, black, steel gray	black	perfect in one direction	submetallic luster; greasy feel; the "lead" in pencils	**Graphite** C
2.5	7.2–7.6 avg 7.4	silver gray	grayish black	three perfect cleavages at 90° (cubes common)	metallic luster	**Galena** PbS
2.5–3	8.94–8.95 avg 8.94	copper red to brown	copper red	none	metallic luster; malleable	**Native Copper** Cu
1.5–6.5	5.3	reddish gray, black	reddish brown	conchoidal fracture	metallic luster; glittering flakes or massive dull varieties common	**Hematite** Fe$_2$O$_3$
5.6–6	5.1–5.2 avg 5.15	black, dark gray	black	uneven fracture	metallic luster; strongly magnetic	**Magnetite** Fe$_3$O$_4$
6.5	5–5.02 avg 5.01	pale brass yellow	greenish black	conchoidal fracture	striations on faces common	**Pyrite** FeS$_2$

Minerals with a Nonmetallic Luster ◆ Colored Streak

Hardness	Density	Color	Streak	Cleavage/Fracture	Other Properties/Comments	Mineral
1.5–2.5	2.05–2.09 avg 2.06	yellow	white	none	transparent to translucent; resinous to vitreous luster	**Sulfur** S
3.5–4	3.77–3.89 avg 3.83	azure blue	light blue	one perfect	vitreous to dull luster; generally occurs as coatings, masses, or tiny crystals; commonly occurs with malachite	**Azurite** $Cu_3(CO_3)_2(OH)_2$
3.5–4	3.6–4 avg 3.8	bright green	light green	fractures in uneven flat surfaces	vitreous to silky luster; generally occurs as coatings, masses, or tiny crystals; commonly occurs with azurite	**Malachite** $Cu_2CO_3(OH)_2$
3.5–4	3.9–4.2 avg 4.05	light yellow to yellowish brown	pale yellow, brownish white	three perfect cleavages	translucent to transparent; resinous to vitreous luster	**Sphalerite** ZnS

Minerals with a Nonmetallic Luster ◆ White or Faintly Colored Streak ◆ No Apparent Cleavage

Hardness	Density	Color	Streak	Cleavage/Fracture	Other Properties/Comments	Name
6.5–7	3.27–3.37 avg 3.32	olive green to yellow green	white	conchoidal fracture	vitreous luster; commonly occurs as sugar-sized granular aggregates	**Olivine** $(Mg,Fe)_2SiO_4$
7	2.6–2.65 avg 2.62	clear, milky, white, purple, smoky, pink	white	conchoidal fracture	vitreous luster	**Quartz** SiO_2

Minerals with a Nonmetallic Luster ♦ White or Faintly Colored Streak ♦ One or More Cleavages

Hardness	Density	Color	Streak	Cleavage/Fracture	Other Properties/Comments	Name
1	2.7–2.8 avg 2.75	pale green; grayish or brownish white	white to light gray	one perfect cleavage	vitreous to pearly luster; soapy feel	**Talc** $Mg_3Si_4O_{10}(OH)_2$
2	2.3	clear; white	white	one perfect cleavage	vitreous to pearly luster; transparent to translucent	**Gypsum** $CaSO_4 \cdot 2H_2O$
2–2.5	2.77–2.88 avg 2.82	clear, silvery white, brownish silvery white	white	one perfect cleavage	vitreous luster; transparent, flexible, and elastic thin sheets	**Muscovite** $KAl_2(AlSi_3)O_{10}(OH)_2$
2.5	2.17	clear, gray, or red	white	three perfect cleavages at 90°	vitreous luster; dissolves in water; salty taste	**Halite** $NaCl$
2.5–3	2.8–3.4 avg 3.09	brown, brownish black	faint brownish gray	one perfect cleavage	vitreous luster; flexible, and elastic thin sheets	**Biotite** $K(Mg,Fe^{2+})_3AlSi_3O_{10}(OH,F)_2$
3	2.71	clear, white	white	three perfect cleavages not at 90°	vitreous to porcelainic luster; reacts strongly in dilute HCl, clear pieces will show double refraction	**Calcite** $CaCO_3$
4	3.01–3.25 avg 3.13	clear, purple, yellow, green, blue	white	four perfect cleavages often resulting in double pyramid	vitreous luster; transparent to translucent	**Fluorite** CaF_2
6	2.56	white, salmon-pink, green, grayish	white	two cleavages at 90° (one perfect)	porcelainic luster	**Orthoclase** $KAlSi_3O_8$
6.5–7	2.64–2.66 avg 2.65	white	white	two cleavages at 90° (one perfect)	porcelainic luster; some cleavage faces have fine light and dark striations	**Plagioclase Oligoclase** $NaCaAl_4Si_4O_8$
6.5–7	2.68–2.71 avg 2.69	dark gray	white	two cleavages at 90° (one perfect)	porcelainic luster; some cleavage faces have fine light and dark striation (twins)	**Plagioclase Labradorite** $CaNaAl_4Si_4O_8$

PreLab
WORKSHEET

NAME _____ **DATE** _____

You ran a series of tests and found a blue mineral to have a hardness between 3 and 4, a density of 3.83 grams/cm³, a light blue streak, and perfect cleavage in one direction. Use the mineral identification key to name the mineral.

Calculate the density of a mineral whose weight is 128.1 grams and whose volume is 14.3 milliliters. Show your calculation, including units in both your calculation and answer.

Chapter 4

Rock Identification

INTRODUCTION

A **rock** is a coherent, naturally occurring solid consisting of an aggregate of minerals, glass, or organic material. The aggregate is held together by the interlocking nature of the materials comprising the aggregate or by a cement.

Rocks reflect their environment of formation so they provide an historical record of geologic events. Geologists split all rocks into three inter-related groups: igneous, sedimentary, and metamorphic. This classification system is based upon the processes involved in the formation of the rock.

IGNEOUS ROCKS

Igneous rocks form by the cooling and crystallization of molten rock (magma). If magma cools slowly deep within Earth over thousands of years, the atoms have time to arrange themselves into crystals, so the resulting rock will be 100% crystals that are large enough to see with the unaided eye. These rocks are referred to as **intrusive** (cooled inside Earth) or **plutonic** (ancient Greek and Roman god of the underworld) rocks. If magma is extruded out on the surface of Earth, it cools quickly, so it doesn't have time to grow all crystals. These rocks are referred to as **extrusive** or **volcanic** rocks. Volcanic rocks may have some crystals that formed while the magma was deep within Earth, but they will always have a fine-grained matrix in which no crystals are distinguishable with the unaided eye.

The classification of igneous rocks is based upon the chemical composition of the magma and the texture (the size, shape, and arrangement of the grains). The texture is **phaneritic** if there are 100% visible crystals, **aphanitic** if there are no visible crystals,

porphyritic if there are large crystals in a finer grained matrix, and **vitreous** if the texture is like glass.

FIGURE 4.1 ▶ *Phaneritic is the texture of 100% crystals, all visible with the unaided eye. Phaneritic texture is the common texture of plutonic rocks.*

FIGURE 4.2 ▶ *Aphanitic is the texture where no crystals are visible with the unaided eye. Aphanitic texture is the common texture of volcanic rocks.*

Magma compositions are commonly grouped into four subdivisions: silicic, intermediate, mafic, and ultramafic. **Silicic** magmas (also know as **felsic** or

FIGURE 4.3 ▶ *Phorphyritic is a texture where there are larger crystals, called phenocrysts, in a finer-grained groundmass (matrix). The groundmass will be smaller visible crystals if the rock is plutonic and aphanitic or glassy if the rock is volcanic.*

FIGURE 4.4 ▶ *Vitreous is the texture where the rock is glassy in appearance. If the glass is dense the rock is called obsidian and if the glass is frothy the rock is called pumice. Pumice will float in water.*

granitic magma) have a SiO_2 content of about 70% with Na, K, and Al being common cations bonding with the silicon-oxygen tetrahedrons. Na, K, and Al tend to make white, light gray, colorless, or pink minerals so rocks originating from silicic magma will tend to be light colored. The most common minerals are feldspars and quartz. The silicic plutonic rock, granite, generally has some dark minerals, usually biotite or hornblende, but the percent of dark-colored minerals is much less than the percent of light-colored minerals.

Magmas of **intermediate** composition have a SiO_2 content of about 60% and have a mix of the minerals common in both silicic and mafic magmas. Plutonic rocks of intermediate composition tend to contain about 50% light color minerals and 50% dark colored minerals. Volcanic rocks of intermediate composition tend to have a porphyritic texture with some visible crystals (commonly plagioclase) in an aphanitic dark matrix.

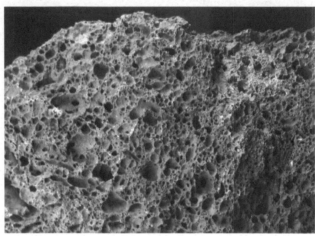

FIGURE 4.5 ▶ *Basalt with numerous vesicles is called scoria.*

Mafic magmas have a SiO_2 content of about 50% with Mg and Fe being common cations bonding with the silicon-oxygen tetrahedrons. Mg and Fe tend to make black, dark gray, or green minerals so rocks originating from mafic magma will tend to be dark colored. The most common minerals will be olivine, pyroxene, amphibole, and Ca-rich plagioclase. Quartz is never present. The mafic volcanic rock, basalt, tends to be very dark gray to black colored and have few, if any, visible crystals. If the basalt has numerous vesicles (air bubbles), it is called scoria.

Ultramafic magmas have a SiO_2 content of about 35%. The ultramafic plutonic rock, peridotite, consists mostly of olivine with some pyroxene so the rock is generally green. Komatiite is a rare ultramafic volcanic rock. Komatiite lava existed prior to about 2.5 Ga

FIGURE 4.6 ▶ *Spinifex is a texture of long thin blades of olivine and shorter blades of pyroxene in an aphanitic matrix.*

when the Earth was much hotter and still capable of erupting lava at temperatures in excess of 1600°C. The diagnostic feature of komatiites is **spinifex texture**, characterized by long thin blades of olivine and shorter blades of pyroxene in an aphanitic matrix.

SEDIMENTARY ROCKS

Sedimentary rocks are classified by the processes that created them. Sedimentary rocks are subdivided into four major groups: clastic, biochemical, chemical, and organic sedimentary rocks.

Detrital Sedimentary Rocks		
Texture (grain size)	Sediment Name	Rock Name
Coarse (over 2 mm)	Gravel (Rounded fragments)	Conglomerate
Coarse (over 2 mm)	Gravel (Angular fragments)	Breccia
Medium (1/16 to 2 mm)	Sand (If abundant feldspar is present the rock is called **Arkose**)	Sandstone
Fine (1/16 to 1/256 mm)	Mud	Siltstone
Very fine (less than 1/256 mm)	Mud	Shale

FIGURE 4.7 ▶ *Classification of clastic (detrital) sedimentary rocks.*

Clastic (detrital) sedimentary rocks form from the fragments of other pre-existing rocks. The pre-existing rocks were exposed at the Earth's surface where they were weathered, transported, deposited, and then buried beneath other sediment, eventually lithifying into solid rock. Clastic sedimentary rocks are classified

on the grain size of the component materials. Grains larger than 2 mm are called **gravel** and the rock is called a **conglomerate** if the **clasts** (rock fragments) are rounded and **breccia** if the clasts are angular. Grains between 2 mm and 1/16 mm are called **sand**, and the rock is called a **sandstone**. Grains between 1/16 mm and 1/256 mm are called **silt**, and the rock is called a **siltstone**. Rocks with grains smaller than 1/256 mm are called **shale**, if the beds have fine laminations and **claystone**, if laminations are not present. The individual grains may or may not be clay minerals but they are too small to see without a microscope.

Chemical and biochemical sedimentary rocks are classified on the minerals present, irrespective of grain size. If the dominant mineral is calcite the rock is called **limestone**. However, if the limestone consists of poorly cemented shell fragments it is called **coquina**; and **chalk**, if the limestone consists

Chemical, Biochemical and Organic Sedimentary Rocks			
Composition	Texture (grain size)	Rock Name	
Calcite, CaCO$_3$	Fine to coarse crystalline	Crystalline Limestone	Biochemical Limestone
Calcite, CaCO$_3$	Fine to coarse crystalline	Travertine	Biochemical Limestone
Calcite, CaCO$_3$	Visible shells and shell fragments loosely cemented	Coquina	Biochemical Limestone
Calcite, CaCO$_3$	Various size shells and shell fragments cemented with calcite cement	Fossiliferous Limestone	Biochemical Limestone
Calcite, CaCO$_3$	Microscopic shells and clay	Chalk	Biochemical Limestone
Quartz, SiO$_2$	Very fine crystalline	Chert	
Gypsum CaSO$_4$•2H$_2$O	Fine to coarse crystalline	Rock Gypsum	
Halite, NaCl	Fine to coarse crystalline	Rock Salt	
Altered plant fragments	Fine-grained organic matter	Bituminous Coal	

FIGURE 4.8 ▶ *Classification of chemical, biochemical, and organic sedimentary rocks.*

of fine-grained, white, calcite. Calcite effervesces in dilute HCl, so this is a good test to identify limestone. If the dominant mineral is dolomite, the rock is called **dolostone**. If the dominant mineral is microcrystalline quartz, the rock is called **chert**. If the dominant mineral is gypsum, the rock is called **rock gypsum**. If the dominant mineral is halite, the rock is called **rock salt**.

Organic sedimentary rocks consist of compressed organic matter. When peat is lithified, it becomes **coal**. The deeper the coal is buried the higher the temperature and pressure it is subjected to, the harder the coal. Coal is generally black and shiny but low grade coal may still have a peat-like appearance.

METAMORPHIC ROCKS

Metamorphism is the solid state recrystallization of minerals due to changes in temperature, pressure, and/or chemically active fluids. Metamorphic rocks are rocks that have undergone metamorphism. Pressure, also called stress, is a force divided by the area of which the force is applied. The stress can be equal in all directions, called non-directed or **lithostatic stress**, or stronger in one direction, called **directed stress**.

Directed stress results in minerals growing longer in the direction perpendicular to the maximum stress so the minerals throughout the rock are all aligned. The texture that results from aligned minerals in a metamorphic rock is called **foliation** (figure 4.9).

Foliation is especially prominent when micas are present.

Under lithostatic stress minerals tend to grow in equant shapes creating triple junctions (120° boundaries between any three adjacent minerals; figure 4.10). Foliation, or the lack thereof, is the primary basis for the classification of metamorphic rocks.

Foliated metamorphic rocks are subdivided into slate, phyllite, schist, and gneiss. **Slate** is a low-grade metamorphic rock with microscopic micas. The micas impart a slight sheen to the rock. In **phyllite**, the micas have grown almost large enough to see with the unaided eye. The larger micas impart a higher sheen then is present in slates. A **schist** is a rock with visible micas. It is common for schists to have abundant micas. At the highest grades of metamorphism, feldspars will begin to form. The resulting rock, **gneiss** will have alternating bands of light minerals, largely quartz and feldspars, and dark minerals, micas and/or amphiboles; so the defining characteristic of gneiss is that it is compositionally banded (Figure 4.9 is a gneiss).

Non-foliated metamorphic rocks are called **hornfels** if the minerals are too small to see and **granofels** if the minerals are large enough to see. There are several common granofels: metamorphosed limestone is called **marble** and metamorphosed quartz-rich sandstone is called **quartzite**.

FIGURE 4.9 ▶ *The above rock has a well developed foliation defined by layers of micas.*

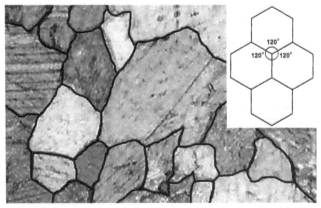

FIGURE 4.10 ▶ *In this photomicrograph (view through a microscope) of marble note that the grains tend to be equant and the grain boundaries tend towards 120°.*

CLASSIFICATION OF IGNEOUS ROCKS

Chemical Composition		Granitic (Felsic)	Andesitic (Intermediate)	Basaltic (Mafic)	Ultramafic
Dominant Minerals		Quartz Potassium feldspar Sodium-rich plagioclase feldspar	Amphibole Sodium- and calcium-rich plagioclase feldspar	Pyroxene Calcium-rich plagioclase feldspar	Olivine Pyroxene
T E X T U R E	Phaneritic (coarse-grained)	Granite	Diorite	Gabbro	Peridotite
	Aphanitic (fine-grained)	Rhyolite	Andesite	Basalt	Komatiite (rare)
	Porphyritic	"Porphyritic" precedes any of the above names whenever there are appreciable phenocrysts			Uncommon
	Glassy	Obsidian (compact glass) Pumice (frothy glass)			
Rock Color (based on % of dark minerals)		0% to 25%	25% to 45%	45% to 85%	85% to 100%

FIGURE 4.11

CLASSIFICATION OF METAMORPHIC ROCKS

Rock Name	Texture		Grain Size	Comments	Parent Rock
Slate	**Increasing Metamorphism**	**Foliated**	Very fine	Excellent rock cleavage, smooth dull surfaces	Shale, mudstone, or siltstone
Phyllite			Fine	Breaks along wavey surfaces, glossy sheen	Slate
Schist			Medium to Coarse	Micaceous minerals dominate, scaly foliation	Phyllite
Gneiss			Medium to Coarse	Compositional banding due to segregation of minerals	Schist, granite, or volcanic rocks
Marble		**Nonfoliated**	Medium to coarse	Interlocking calcite or dolomite grains	Limestone, dolostone
Quartzite			Medium to coarse	Fused quartz grains, massive, very hard	Quartz sandstone
Anthracite			Fine	Shiny black organic rock that may exhibit conchoidal fracture	Bituminous coal

FIGURE 4.12

NAME _____ **DATE** _____

Name the silicic plutonic rock.

Name the clastic sedimentary rock with a grain size of 1/16-2 millimeters.

What is metamorphosed limestone called?

Chapter 5

Soils and Groundwater

SOIL

Soil is the unconsolidated material on the surface of Earth that supports the growth of plants. Soil consists of about 50% mineral matter and **humus** (partially decayed organic matter) and about 50% pore spaces which contain air and water. All four ingredients must be present to support the growth of plants.

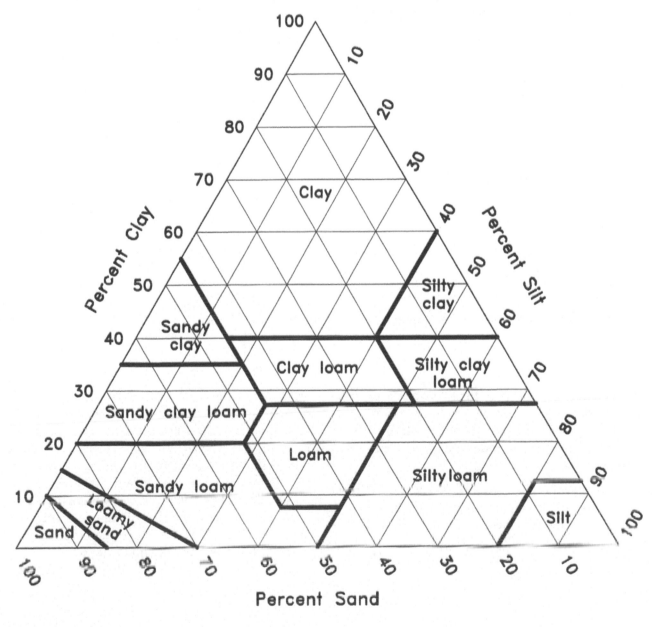

FIGURE 5.1 ▶ *Soil Textural Classification Chart.*

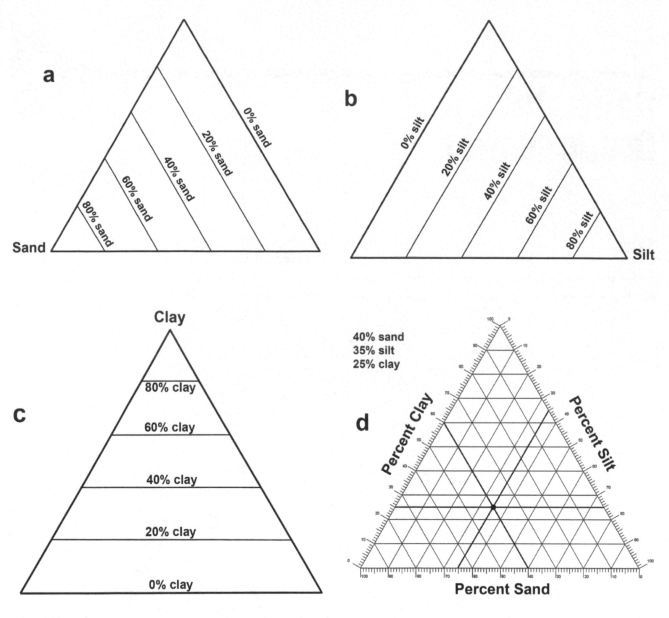

FIGURE 5.2 ▶ *How to read a triangular graph. See text for discussion.*

Soil is classified based upon the relative proportions of sand, silt, and clay-sized particles. **Sand** consists of grains 1/16 mm–2 mm in diameter, **silt** consists of grains 1/16 mm–1/256 mm in diameter, and **clay** consists of grains < 1/256 mm.

A soil classification chart is a three-sided graph drawn on an equilateral triangle. A triangular graph allows you to plot three variables (sand, silt, and clay) that sum to a constant (100%). The base of the triangle shows the percent sand from 0% sand on the right to 100% sand on the right (figure 5.2a). The right side of the triangle shows the percent silt from 0% silt at the top of the triangle to 100% silt in the lower right corner of the triangle (figure 5.2b). The left side of the triangle shows the percent clay from 0% clay in the lower left corner of the triangle to 100% clay at the top of the triangle (figure 5.2c). The dot in Figure 5.2d shows the location of a soil sample that is 40% sand, 35% silt, and 25% clay. Note that the percents of sand, silt, and clay must add up to 100%.

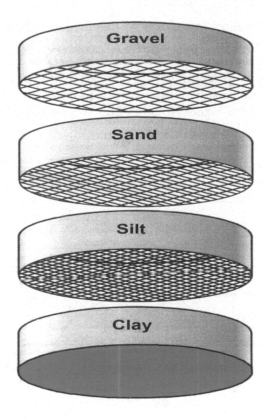

FIGURE 5.3 ▶ *Sieves for sorting grain sizes.*

CLASSIFYING AN UNKNOWN SOIL SAMPLE

A series of sieves are used to separate a soil sample into sand, silt, and clay-sized particles. A sieve is a container with a crimped wire mesh bottom such that particles smaller than the mesh spacing will fall through the mesh and those larger than the mesh spacing will remain in the container. By using a series of sieves with different mesh spacing, it is possible to separate a soil sample into sand, silt, and clay sized particles.

Begin with a known volume of soil sample.

Stack the sieves so that the screen with the largest openings is on top. Each sieve has progressively smaller screen openings until the bottom sieve has a solid bottom without any screen.

Pour the soil sample into the top sieve and secure the lid tightly.

Shake the sieves side to side for at least one minute.

Remove the lid and the top sieve from the stack. All the particles in this sieve are larger than 2 mm, so are not considered to be soil particles. Set them aside.

Fasten the lid to the remaining stack and shake the sieves side to side for about 2 minutes.

Remove the lid and separate the remaining sieves. Use a funnel to pour the material in the top sieve into a graduated cylinder. Gently tap the side of the cylinder until the material achieves a level surface. Do not pack down the soil because this will reduce the volume by reducing the size of the spaces between the grains. Record the volume of sand in milliliters.

Repeat for silt and clay.

Add the volume of sand, silt, and clay together to get the total volume of the soil sample.

Determine the percent sand, silt, and clay. The percent sand is the volume of sand divided by the total volume of soil times 100%. Repeat for silt and clay.

Determine the textural classification of your soil sample by plotting the percentages of sand, silt, and clay on the "Soil Textural Classification Chart in figure 5.1."

GROUNDWATER

During a rainstorm some of the water soaks (infiltrates) into the ground. Water that infiltrates into the ground travels downward until it reaches a zone, called the **phreatic zone** or the **zone of saturation,** where 100% of the pore spaces are filled with water. The water in the phreatic zone is called

Porosity in sediment

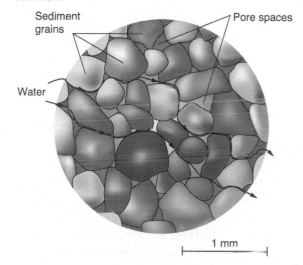

FIGURE 5.4 ▶ *Pore space is the space between grains.*

groundwater. The top of the phreatic zone is called the **water table.**

If you look closely at unconsolidated material, you will notice that the grains do not fit tightly together. The spaces between the grains are called **pore spaces.**

$$porosity = \frac{\textbf{volume of pore space}}{\textbf{volume of sediment or rock}}$$

The volume occupied by pore spaces is called the **porosity** and is usually stated as a percent.

The porosity is influenced by grain shape, arrangement, and sorting (distribution of grain sizes). A well sorted sediment where the grains are mostly the same size will have a higher porosity than a poorly sorted sediment where the smaller grains can lodge in the spaces between the larger grains, thus reducing the porosity. Well sorted sediments typically have porosities of about 50%. Typical porosities in rocks range from about 1% to about 35%.

MEASURING POROSITY

Place 100 ml of sediment into a beaker.

Place 100 ml of water into a graduated cylinder.

Slowly pour the water into the beaker until the sand is entirely saturated and the water level in the beaker just equals the level of the surface of the sediment.

Record the volume of water you added (the original 100 ml minus the amount of water left in the graduated cylinder):

The porosity of the sediment is the volume of water you added to the beaker divided by the volume of the original sediment (100 ml).

GROUNDWATER FLOW

Pore spaces are generally connected to each other, so fluids can move through the pore spaces. The ability of fluids to move through the pore spaces depends upon the size and connectedness of the pore spaces, a property called **permeability** or **hydraulic conductivity.** The porosity and permeability of the sediments (or rocks) control the velocity of groundwater movement.

The elevation of the water table (called **hydraulic head**) changes from location to location. If the elevation of the hydraulic head changes between two locations, the groundwater will tend to flow under the influence of gravity from the area where the water table is at a higher elevation into an area where the water table is at a lower elevation. The direction of groundwater flow can be determined by contouring the elevation of the water table throughout an area. Groundwater flows down slope (hydraulic gradient) perpendicular to the contours at any given location.

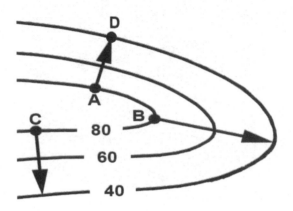

FIGURE 5.5 ▶ *A contour map showing the elevation of the water table above sea level. The direction of groundwater flow is shown for three locations: A, B, and C. Note that the groundwater flows (arrows) down slope from each location in a direction that is perpendicular to the contours in that area.*

CALCULATING THE HYDRAULIC GRADIENT

The hydraulic gradient between two wells is the elevation of the water table at the first well minus the elevation of the water table at the second well divided by the distance between the first and second wells.

$$\text{Hydraulic gradient} = \frac{\begin{array}{c}\textbf{elevation in 1st well} -\\ \textbf{elevation in 2nd well}\end{array}}{\begin{array}{c}\textbf{horizontal distance between}\\ \textbf{the wells}\end{array}}$$

For example, if the distance between well A and well D in figure 5.5 is 2000 feet then the hydraulic gradient is (80 feet − 40 feet) / 2000 feet = 0.020

TABLE 5.1 ▶ *Typical porosity, n, and hydraulic conductivity, K, for coarse sand and silt.*

sediment	n	K (feet/day)
coarse sand	0.35	147
silt	0.40	0.27

CALCULATING GROUND WATER VELOCITY

The groundwater velocity in feet/day is the hydraulic gradient times the hydraulic conductivity divided by the porosity.

$$\text{Groundwater velocity} = \frac{\text{Hydraulic gradient} \times \text{hydraulic conductivity}}{\text{porosity}}$$

For example, a contaminant in well A of figure 5.5 moving through a coarse sand aquifer towards well D would move at a velocity of (0.020 x 147ft/day)/0.35 = 8.4 feet/day.

NAME _____ DATE _____

What is the classification of a soil that is 35% sand, 45% silt, and 20% clay?

You found that a sandstone was capable of holding 18 ml of water. Calculate the porosity of the sandstone if the volume of the sandstone is 100 ml. Show your calculation, including units.

Calculate the hydraulic gradient between two wells that are 1000 feet apart, given that the elevation of the water table in the first well is 980 feet and the elevation of the water table in the second well is 970 feet. Show your calculation, including units.

Chapter 6

Streams

INTRODUCTION

The term **stream** is a generic term used to denote any channelized flow regardless of size. Large streams with many tributaries are generally referred to as **rivers**. A **drainage basin** is the area drained by a stream system. Each basin is bounded by a divide beyond which water is drained by another stream system. A **divide** is a relatively high area that separates the drainage of two streams. It is common practice to measure the volume of water flowing in streams throughout a drainage basin. **Discharge** is the volume of water passing a given point during a specific interval of time. Discharge is commonly measured in cubic feet (or meters) per second.

Velocity is not uniform throughout a stream channel. In a straight section of stream channel, the velocity is greatest in the center of the channel just below the surface, away from the frictional drag of the channel sides and bottom.

As the channel curves around a **meander** (loop-like bend in a stream channel), centrifugal force comes into play, so the zone of maximum velocity shifts to the outside of the bend and a zone of minimum velocity exists along the inside of the bend. The higher water velocity along the outside bank of the meander erodes the bank, creating a **cutbank**. The slower water along the inside of the meander loop results in a deposit called a **point bar**.

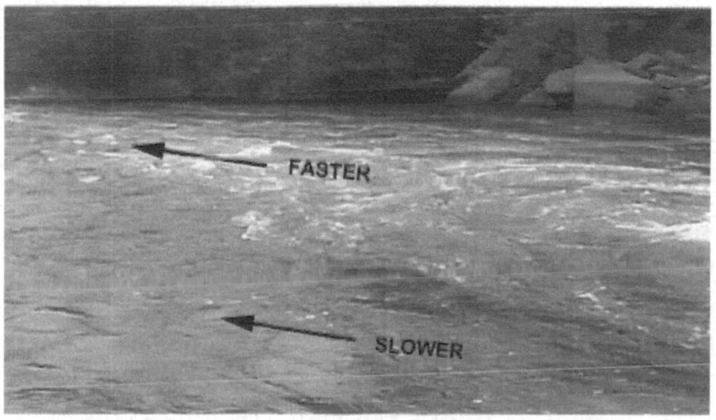

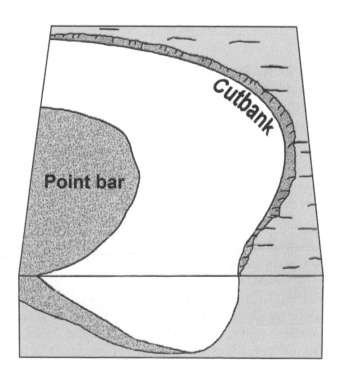

FIGURE 6.1 ▶ *Note in this stream that the faster water is near the center of the stream and the slower water is closer to the stream banks.*

FIGURE 6.2 ▶ *In a stream meander, there is erosion along the cutbank side of the stream and deposition along the point bar side of the stream.*

DETERMINING STREAM DISCHARGE

Stream discharge is the volume of water passing a given point over a specific interval of time. Discharge is commonly measured in feet³/second (or meters³/second). To calculate stream discharge, you must measure the area of the stream and the velocity of the flow of water in the stream.

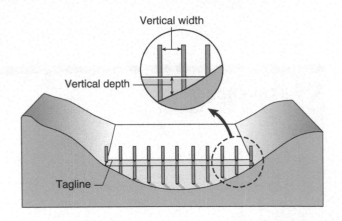

FIGURE 6.4 ▶ *Stream verticals showing vertical width and vertical depth at vertical center. The average velocity is measured along the vertical center.*

FIGURE 6.3 ▶ *A wading rod with a rotating sensor attached to the bottom and a display meter connected to the top is used to measure water velocity.*

Velocity is measured by a sensor attached to the lower part of a wading rod. The sensor blades rotate in the current and the velocity is read on a digital display meter (figure 6.3).

Water depth and velocity varies across a stream. Water velocity also varies from the surface of a stream down to the stream bed. To determine the total discharge of a stream, we are going to use the *velocity-area method*. In this method, stream velocity and water depth measurements are taken at a minimum of ten locations (called **verticals**) along a transect (called a **tagline**) perpendicular to the stream (see figure 6.4).

If a stream is 300 feet wide, dividing the 300 foot wide stream into 10 equal verticals yields a vertical width of 30 feet. The velocity is taken at the center of

each vertical. For example, the center of the first vertical which starts at the stream bank and ends 30 feet into the stream is 15 feet.

The velocity of the water varies from the water's surface down to the stream bed, so velocity readings must be taken at various depths (called depth fractions). The **depth fraction** is the fraction of the stream depth in a vertical. For example, 0.1 is 10% of the stream depth in a vertical, so if the stream is 10 feet deep in the vertical being measured, 0.1 would be one foot below the surface of the water, 0.2 would be two feet below the surface, and so on. Readings are not taken for the surface and bottom of the stream since the readings would be inaccurate.

Rather than measure the velocity nine different times for each vertical, experience has shown that you can take one reading at the depth of the average velocity in a vertical but what depth does the average velocity correspond to? In the Prelab you will determine the depth of the average velocity. The depth of the average velocity occurs at the same depth fraction for most streams so you only need to take one velocity reading for each vertical. This is referred to as the 6/10ths rule.

The stream discharge for each vertical is the vertical area times the average velocity. The area is the vertical width times the vertical depth. The total discharge of the stream is then calculated by adding the discharge from each vertical.

FLOODPLAINS AND TERRACES

A stream **floodplain** is a relatively flat area immediately adjacent to the stream channel (figure 6.5). When stream discharge exceeds the capacity of the channel to contain the water, the water flows out across the floodplain. Relatively flat areas that lie above the floodplain are called **terraces**. Stream terraces often are former floodplains when the stream carried significantly more water, such as when the glaciers were melting at the end of the last ice age.

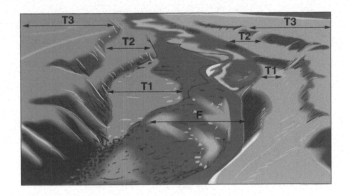

FIGURE 6.5 ▶ *A diagram showing a floodplain (F) and several terraces. T1, T2, and T3 are successively higher stream terraces.*

PreLab

NAME _____ DATE _____

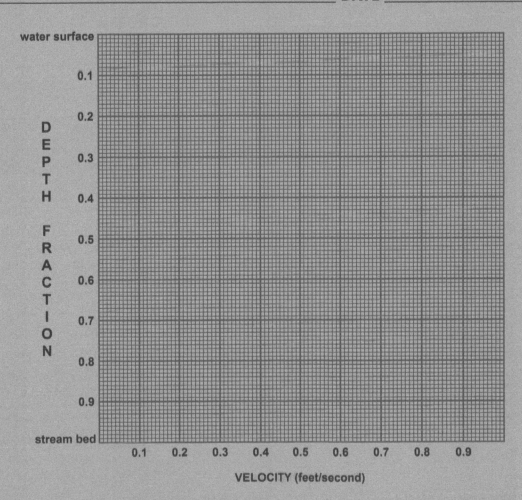

Plot the following data on the graph above.

Depth Fraction	Velocity (ft./sec.)
0.1	0.84
0.2	0.80
0.3	0.76
0.4	0.70
0.5	0.64
0.6	0.58
0.7	0.48
0.8	0.34
0.9	0.10
Sum of velocities	
Average velocity =	

To determine the depth of the average velocity, add the nine velocities together and record your answer on the table. To calculate the average velocity, divide the total of the nine velocities by nine. On the Depth Fraction – Velocity graph above draw a vertical line upward from the average velocity to where it intersects your curve; then draw a horizontal line to the depth fraction that most closely corresponds to the average velocity of the stream.

What depth fraction corresponds to the average velocity?

Chapter 7

Glaciation

INTRODUCTION

A **glacier** is a mass of ice formed from recrystallized snow that deforms plastically under its own weight. If year after year more snow falls than melts, an entire mountain range or even a continent can be buried. As the snow gets deeper, the increased pressure turns the snow into ice. The ice at the bottom of a glacier is under a lot of pressure caused by the weight of the overlying ice. Ice is not very strong, so the ice deforms like putty when you squeeze it.

Glaciers are divided into two major categories based on where they form: alpine glaciers and continental glaciers. **Alpine glaciers** (also known as valley glaciers) are long narrow rivers of ice that accumulate in high mountains

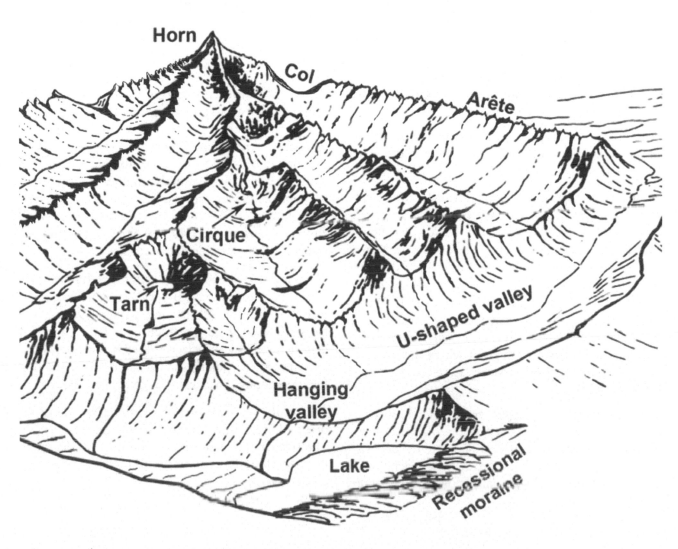

FIGURE 7.1 ▶ *Erosional landforms created by alpine glaciers. See text for discussion.*

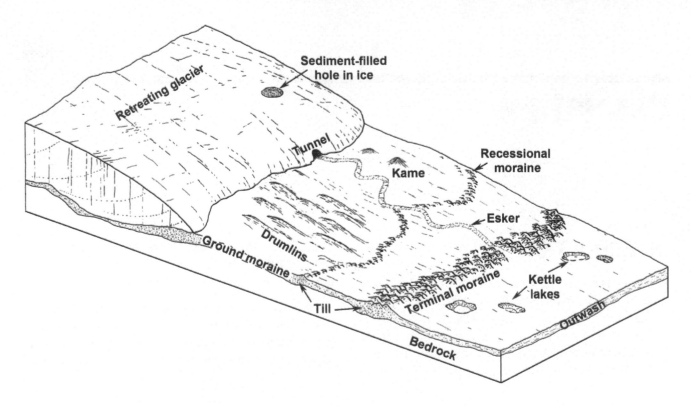

FIGURE 7.2 ▶ *Depositional landforms created by glaciers.*

and flow down pre-existing stream valleys. **Continental glaciers** are large ice sheets that cover continents.

Glaciers are effective agents of erosion. Most of the erosional landforms we will examine in this lab are associated with alpine glaciation.

An **arête** is a jagged sawtooth ridge separating two glacial valleys. A **horn** is a jagged pinnacle peak where several arêtes coalesce. A **cirque** is a bowl-shaped area at head of an alpine glacial valley. A **tarn** is a lake in a cirque. The amount of erosion depends upon the thickness of the ice so large glaciers erode deeper valleys than smaller tributary glaciers. When the glaciers melt, the floor of a smaller tributary glacial valley is at a higher elevation than the floor of a larger glacial valley. At the point where the two glaciers meet, the tributary valley is referred to as a **hanging valley**.

The material eroded by a glacier is carried in suspension in the ice and deposited at the margins of the glacier. **Till** is unsorted, unstratified material deposited directly by a glacier. **Outwash**, a glaciofluvial deposit, is deposited by glacial meltwater so is better

sorted. **Moraines** are ridges or mounds of till. Moraines that are deposited as ridges of till at the end of a glacier during a period of recession, are referred to as **end** or **recessional** moraines. The farthest end moraine that marks the maximum advance of a glacier is called the **terminal** moraine. Moraines that are deposited along the sides of a glacier are referred to as **lateral** moraines, if they are associated with alpine glaciers and **interlobate** moraines, if they are associated with continental glaciers.

There are a few depositional landforms that are only associated with continental glaciers: ground moraines, drumlins, kettles, eskers, and kames. Continental glaciers are so large that there may be a continual deposit of till as the glacier recedes. These **ground** moraines are irregularly shaped and result in a rolling hill topography consisting of high and low areas.

A **drumlin** is a streamlined asymmetrical hill of till oriented in the direction of movement of the glacier. The steeper side is in the direction the glacier came from and the gentler slope points in the direction the glacier moved.

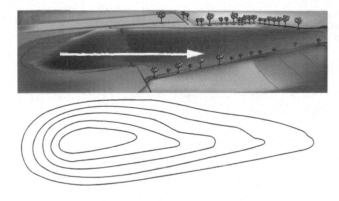

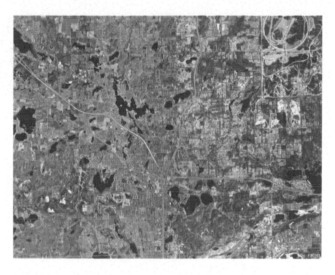

FIGURE 7.3 ▶ *A drumlin showing the classic streamlined asymmetrical shape. The arrow points in the direction of glacial advance. The lower diagram shows the typical shape of a drumlin as displayed by topographic contours.*

FIGURE 7.4 ▶ *Numerous kettle lakes in the vicinity of Brighton, Michigan.*

A **kettle** is a depression formed when a block of ice is buried or partially buried by outwash and then later melts. If the depression is filled with water it is referred to as a **kettle lake**.

Continental glaciers commonly have streams flowing in tunnels along the bottom of the glacier. The tunnel may become clogged with sediment. After the glacier melts, some of these glacial stream tunnel deposits may be preserved as an **esker**, a long sinuous ridge of melt water stream deposits from channels flowing along the bottom of a melting glacier.

Depressions on the surface of continental glaciers may become filled with water-born sediment. After the glacier melts, some of these deposits may be preserved as one of the several forms of kames. In this case, the **kame** is an irregularly-shaped hill of glaciofluvial sediment.

FIGURE 7.5 ▶ *A classic sinuous esker.*

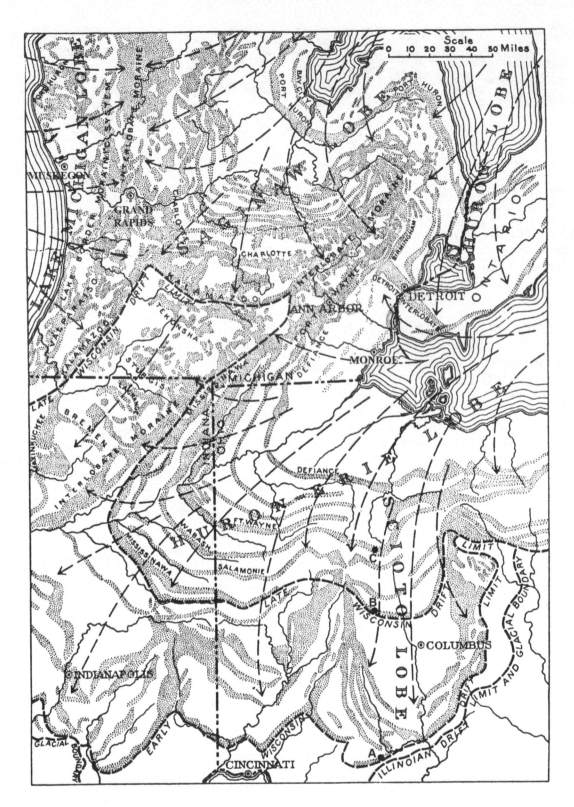

FIGURE 7.6 ▶ *Glacial moraines of southern Michigan and portions of adjacent states. The stippled areas are moraines. The direction of ice movement of the different lobes is indicated by arrows and the limits of the ice at different stages by heavy dashed lines. After Leverett and Taylor, 1915, The Pleistocene of Indiana and Michigan and the history of the Great Lakes: U.S.G.S. Monograph LIII, Plate V, p. 62.*

NAME _____ DATE _____

FIGURE 7.7 ▶ *Alpine glaciation in the Bitterroot Mountains along the Idaho—Montana border.*

What is the glacial erosional landform at **A** in figure 7.7?

What is the name given to the head of the glacial valley in the general area of **B** in figure 7.7?

What is the glacial erosional landform at **C** in figure 7.7?

Coastal Processes

Chapter 8

TIDES

Tides are periodic changes in the elevation of the ocean surface at any given location due to gravitational influences of the moon and sun and to centrifugal forces. The gravity of the moon pulls the ocean's water towards the moon creating a high tide. The centrifugal forces generated by the Earth-Moon system result in another high tide on the side of the Earth opposite of the moon. Based upon this information, you would expect there to be two high tides and two low tides each tidal day with each being about equal height (called **semidiurnal tides**). A **tidal day** is 24 hours 50 minutes because the moon advances 50 minutes each day in its orbit around Earth. The size and shape of ocean basins and coastlines vary from location to location and can affect the tides; so at any given location there can be one high tide and one low tide per day of about equal height (called **diurnal tides**), or two high tides and two low tides but of significantly different heights (called **mixed tides**).

The tide height is referenced to the **mean lower low water** (MLLW), which is the average daily height of the lowest local tide as measured over several years. Table 8.2 shows tide data from Bar Harbor, Maine.

TABLE 8.1 ▶ *Lunar information for a few days in April 2006.*

DATE	TIME	
4/10/06	4:19 pm	moon rise
4/10/06	10:44 pm	zenith
4/11/06	5:09 am	moon set
4/11/06	5:22 pm	moon rise
4/11/06	11:23 pm	zenith
4/12/06	5:25 am	moon set

TABLE 8.2 ▶ *Tide data from Bar Harbor, Maine.*

DATE	TIME	MLLW (ft)
4/10/06	10:00 am	10
4/10/06	12:00 pm	6
4/10/06	4:00 pm	1
4/10/06	7:00 pm	6
4/10/06	10:00 pm	10
4/11/06	1:00 am	6
4/11/06	4:00 am	1
4/11/06	7:00 am	4
4/11/06	10:00 am	10
4/11/06	1:00 pm	6
4/11/06	4:00 pm	1
4/11/06	7:00 pm	4
4/11/06	10:00 pm	10
4/12/06	2:00 am	4
4/12/06	5:00 am	1

LONGSHORE CURRENTS

As waves approach the shore they are bent (**wave refraction**) and tend to become subparallel to the shoreline. The refraction occurs because the part of the wave nearest the shore is in shallower water, so the friction between the wave and the ocean floor slows down the advance of the wave. The shallower the water, the more friction, the slower the wave will advance.

Although waves are refracted, most waves still reach the shore at an angle to the shoreline. This generates a current parallel to the shoreline, called a **longshore current.** The longshore current transports sediments along the beach, a process called **longshore drift.** You can determine the direction of the longshore current by looking at the overall shape of the sedimentary deposits along the beach (figure 8.1).

Barriers are often built to control the migration and deposition of beach sand. A **groin** is a barrier build at a right angle to the beach to trap sand that is moving parallel to the shore via longshore currents. Generally, a series of groins are built to control sand migration along a section of beach. **Jetties** are parallel dikes on either side of a river to keep sand from accumulating and blocking a harbor. A **breakwater** is an offshore barrier that protects a beach or harbor from the full force of the waves. Longshore drift results in sediments accumulating behind groins, jetties, and breakwaters.

WAVES

Most waves are set in motion when friction from wind begins rotating water particles in circular orbits (figure 8.2). We observe a wave crest at the top of the circular orbit and a wave trough at the bottom of the circular orbit. The wave height is the vertical distance between the wave crest and trough. Beneath the surface, in deep water the circular orbits of water molecules become progressively smaller with depth. At a depth equal to about one half the **wavelength** (horizontal distance between successive wave crests) the circular motion of the water has decreased to zero. In deep water, where the depth is greater than one half the wavelength, the velocity, V, of a wave depends upon the **wave period** (the time interval between successive crests at a stationary point), T, and the wavelength, L.

$$\text{Velocity} = \frac{\text{wavelength}}{\text{wave period}}$$

As a wave approaches the shore and the depth of water becomes less than one half the deep water wavelength, the circular orbit of the water molecules touch the ocean floor. Friction with the ocean floor causes changes in the circular motion of the water. At a depth equal to about 1/20th of the deep water wavelength, the top of the wave begins to fall forward and the wave breaks. In the surf zone, where waves are breaking and releasing energy, a significant amount of water is transported toward the shoreline.

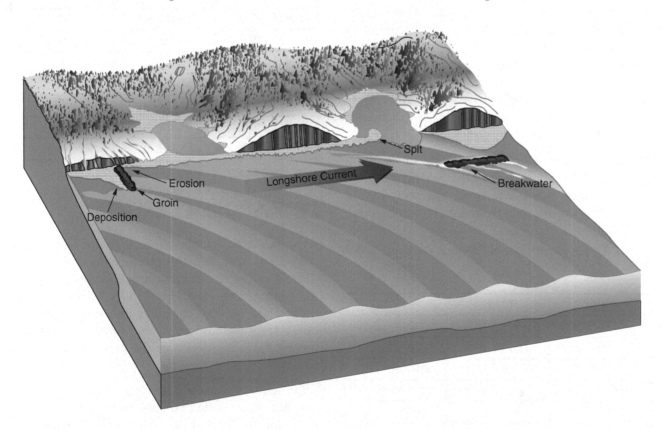

FIGURE 8.1 ▶ *Illustration showing the relationship between the direction of the longshore current and the geometry of longshore drift deposits.*

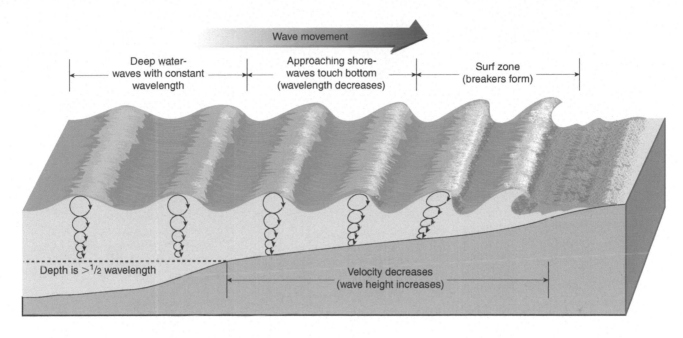

Wave movement

Deep water-
waves with constant
wavelength

Approaching shore-
waves touch bottom
(wavelength decreases)

Surf zone
(breakers form)

Depth is >½ wavelength

Velocity decreases
(wave height increases)

FIGURE 8.2 ▶ *Waves in deep and shallow water.*

SEA LEVEL RISE

The glaciers on Greenland and Antarctica are melting due to global warming. The sea level rises as a result of the water from the melting glaciers. The rising sea level results in the oceans transgressing (advancing) across the land. Many coastal cities of today may be below sea level in the future.

TABLE 8.3 ▶ *Data on Greenland and Antarctica ice and ocean water.*

Greenland	
Average thickness of ice	4950 ft
Surface area of ice	672,000 mi^2
Volume of ice	630,000 mi^3
Antarctica	
Average thickness of ice	6600 ft
Surface area of ice	4,800,000 mi^2
Volume of ice	6,000,000 mi^3
Oceans	
Average depth of water	12,382 ft
Surface area of oceans	139,000,000 mi^2
Volume of water	326,000,000 mi^3

NAME _____ DATE _____

What would be the velocity in miles/hour of deep water waves with a wavelength of 100 feet and a wave period of 10 seconds? Show your calculation, including units.

What would be the wavelength, in feet, of deep water waves that have a period of 10 seconds and a velocity of 10 feet/second? Show your calculation, including units.

At what depth, in feet, would a wave begin to break if it had a wave period of 10 seconds and a velocity of 10 feet/second? Show your calculation, including units.

If a tsunami (an earthquake generated wave) had a wavelength of 100 miles and a period of 15 minutes, what would be its velocity in miles/hour? Note: 15 minutes is 0.25 hours. Show your calculation, including units.

Chapter 9

Earthquakes and Plate Tectonics

EARTHQUAKES

Earthquakes are vibrations of the earth caused by the rupture and sudden movement of rocks strained beyond their elastic limits. Fractures in the Earth along which movement occurs are called **faults**. Along some faults movement occurs easily, so stress does not build up. Along other faults the friction is greater, so the fault will not move until there is a large build up of stress. Rocks on either side of a "locked fault" bend as stress builds up. When the stress exceeds the friction between the two sides of the fault, the rocks suddenly snap back to their pre-stress orientation, with an abrupt displacement along the fault. Vibrations associated with this movement are called **seismic waves**.

The vibrations can be subdivided into those that vibrate in the same direction they are propagating (**Primary waves** or **P-waves**) and those that vibrate perpendicular to the direction they are propagating (**Secondary waves** or **S-waves**). P-waves propagate through the crust of the Earth at about 6 kilometers/second; whereas S-waves travel through the crust at about 4 kilometers/second. Seismic waves propagate outward in all directions from the point of rupture within the earth (called the **focus**), but since P and S waves travel at different velocities, they arrive at distant seismic stations at different times. Since the propagation velocity is known for both P and S waves it is possible to determine the location of the earthquake **epicenter** (the geographic point on the surface of the Earth directly above the focus) by noting the time lag between the arrival of the first P-wave and the first S-wave.

DETERMINING AN EARTHQUAKE EPICENTER

Figure 9.1 contains seismograms from Billings, Montana; Denver, Colorado; and Phoenix, Arizona. Each seismogram records the same earthquake.

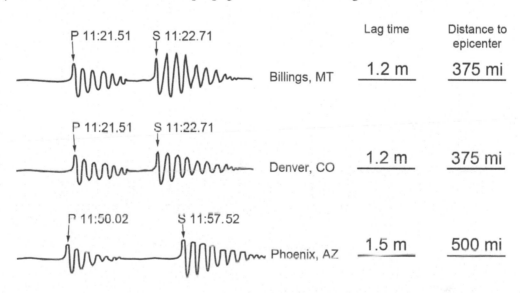

	Lag time	Distance to epicenter
P 11:21.51 S 11:22.71 Billings, MT	1.2 m	375 mi
P 11:21.51 S 11:22.71 Denver, CO	1.2 m	375 mi
P 11:50.02 S 11:57.52 Phoenix, AZ	1.5 m	500 mi

FIGURE 9.1 ▶ *Seismograms recorded at Billings, Montana, Denver, Colorado, and Phoenix, Arizona. The distance to the earthquake epicenter is determined by plotting the lag time on the diagram of travel time curves for P-waves and S-waves (figure 9.2).*

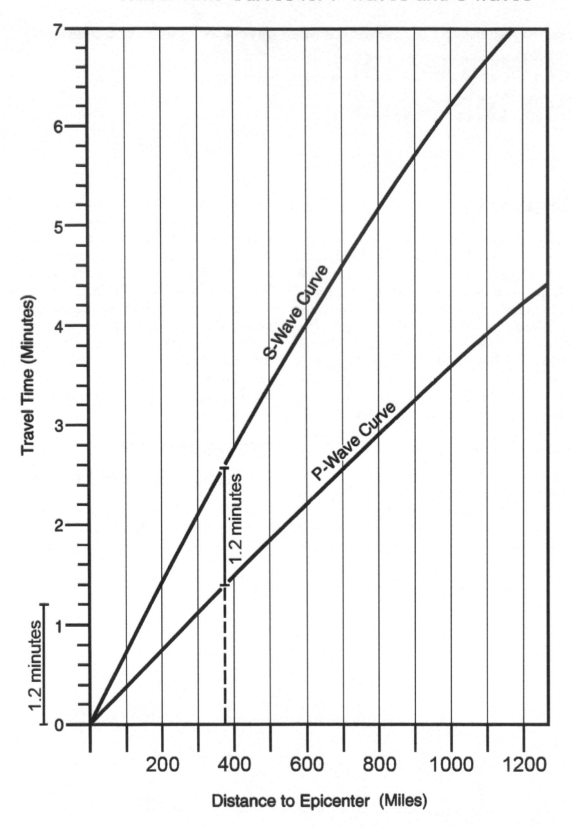

FIGURE 9.2 ▶ *Travel time curves for P waves and S waves.*

The P-wave and S-wave first arrivals on each seismogram are labeled (P and S) and are marked by a dramatic increase in amplitude. The times of the first arrivals are given in hours, minutes, and decimal fractions of minutes (*not* minutes and seconds). To determine the **lag time**, subtract the P-wave first arrival time from the S-wave first arrival time. To determine what distance from the epicenter this lag time corresponds to, use the "Travel Time Curves for P waves and S waves" (figure 9.2).

For example, if the first arrival of the P-wave was at 11:21.51 UT (Universal Time—formerly Greenwich Mean Time) and the first arrival of the S-wave was at 11:22.71 UT, the lag time would be 1.2 minutes. Use the travel time (figure 9.2) curve to determine a 1.2 minute interval between the P-wave curve and the S-wave curve. Note that the time interval is *not* plotted as 1.2 minutes along the vertical axis, the time interval line is plotted as a separation between the P-wave curve and the S-wave curve that corresponds to the length represented by 1.2 minutes. Projecting downward to the horizontal axis reveals that a 1.2 minute lag time corresponds to a distance of about 375 miles to the epicenter.

The epicenter is determined by using a compass and the scale on the map of the United States (figure 9.3) to plot a circle of radius 375 miles for Billings and Denver and 500 miles for Phoenix. The epicenter of the earthquake is at the intersection of the three circles, in this case, along the Wasatch fault near Salt Lake City, Utah. For shallow focus earthquakes, the three circles will intersect at a point. For deep focus earthquakes, the intersection will not be an exact point but a small triangle (a few millimeters). A triangle larger than a few millimeters implies that either your data or circles are inaccurate.

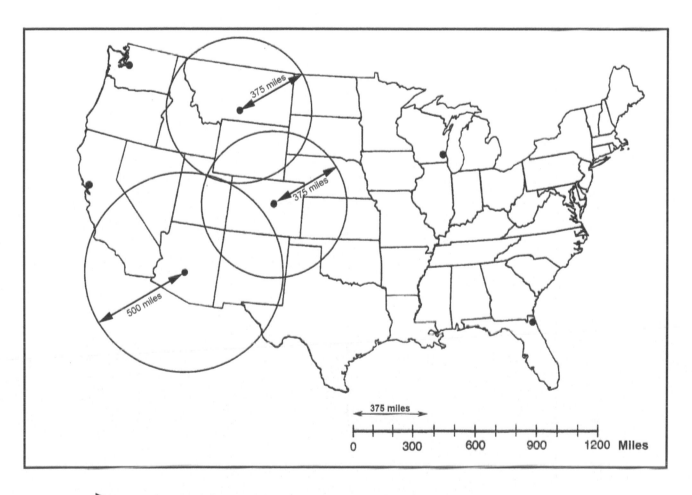

FIGURE 9.3 ▶ *The epicenter of the example earthquake is near Salt Lake City, Utah.*

DETERMINING EARTHQUAKE MAGNITUDE

Charles Richter developed an earthquake magnitude scale based upon the greatest amplitude of the seismic waves and the distance from the epicenter. The seismic wave with the greatest amplitude at any given location can be the P-wave, the S-wave, or surface waves.

The **lag time** is the time interval between the arrival of the first P-wave and the arrival of the first S-wave. The **amplitude** is the height in millimeters. In the example shown in figure 9.5, the lag time is 45 seconds, which corresponds to a distance of 400 kilometers from the epicenter and an amplitude of 300 millimeters. Rather than calculate the magnitude it is easier to use a **nomogram**, which is a graphical representation of a mathematical relationship between three related functions. A nomogram contains three graphs, two with known values (in this case, lag time and amplitude)

plotted on the outside of the nomogram and an unknown value (magnitude) in the center. Draw a line between the two known values (lag time 45 seconds (400 km) and amplitude 300 millimeters). The Richter magnitude is where the line crosses the magnitude scale, in this case 6.9.

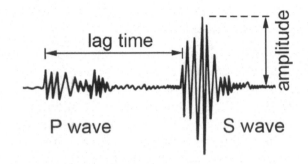

FIGURE 9.4 ▶ *A seismogram showing P and S waves, lag time, and amplitude.*

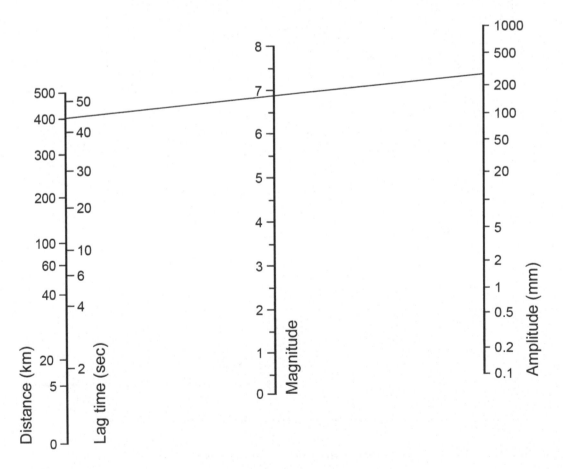

FIGURE 9.5 ▶ *Nomogram for determining Richter earthquake magnitude.*

PLATE TECTONICS

The relatively rigid upper part of the earth (lithosphere) is broken into pieces (plates) and these plates are constantly moving. Two adjacent plates can be moving apart (divergent plate boundary), coming together (convergent plate boundary), or sliding past each other (transform plate boundary). Since there is movement along plate boundaries, they are essentially large scale faults. The majority of earthquakes occur along plate boundaries, so earthquake epicenters can be used to identify plate boundaries. In areas where the lithosphere is being pulled apart, magma moves up through the fractures so active volcanism is common along divergent plate boundaries. Convergent plate boundaries involving subduction can be identified by the presence of an oceanic trench. Convergent plate boundaries will also have volcanically active young mountain ranges or island arcs landward from the trench.

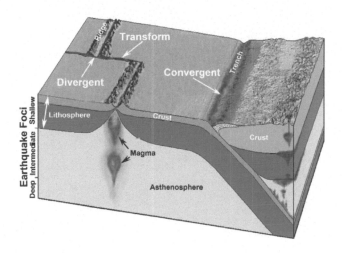

FIGURE 9.6 ▶ *Divergent, convergent, and transform plate boundaries. See text for discussion.*

Oceanic lithosphere averages about 70 kilometers thick and continental lithosphere averages about 150 kilometers thick. Earthquakes only happen in the lithosphere so divergent and transform plate boundaries only have shallow **focus** earthquakes (the place within the earth where slip along a fault begins). If an earthquake focus is deeper than 150 kilometers it implies that the lithosphere is being subducted down into the mantle. This only occurs along convergent plate boundaries. Magma is often generated at depths of 150–250 kilometers along convergent plate boundaries.

NAME _____ **DATE** _____

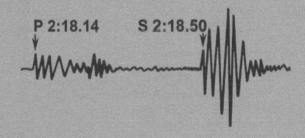

FIGURE 9.7 ▶ *Seismogram showing P and S waves from an earthquake.*

Determine the lag time between the P-wave first arrival and the S-wave first arrival in figure 9.7. Show your calculation, including units.

Determine the distance to the earthquake epicenter using the data from figure 9.7. Hint: use the travel time curves in figure 9.2

Determine the magnitude of the earthquake in figure 9.7. Hint: use the nomogram in figure 9.5.

Chapter 10

Seasons, Solar Radiation, and Standard Time

INTRODUCTION

The amount of solar radiation that reaches the Earth varies with time and location. The variance is largely a result of the Earth's rotation, revolution, and the inclination of Earth's rotation axis to the plane of the Earth's orbit around the sun (plane of the ecliptic). **Rotation** is the spinning of the Earth on an axis that goes through the poles of the Earth. Rotation so profoundly effects us that a **day** is defined as the time it takes the sun to pass from one zenith (the highest point the sun reaches as it travels across the sky) to the next. This interval of time (one **day**) is defined as 24 hours. **Revolution** is Earth's motion around the sun. The path is referred to as the **orbit**. It takes 365¼ days for the Earth to make one complete revolution around the sun; this time interval is referred to as one **year**.

Earth travels in an elliptical orbit around the sun. Earth is closest to the sun (**perihelion**) on January 2–5 (the exact time varies from year to year). At perihelion, Earth is 91,500,000 miles from the sun. Earth is farthest from the sun (**aphelion**) on July 3–7 (the exact time varies from year to year). At aphelion, Earth is 94,500,000 miles from the sun.

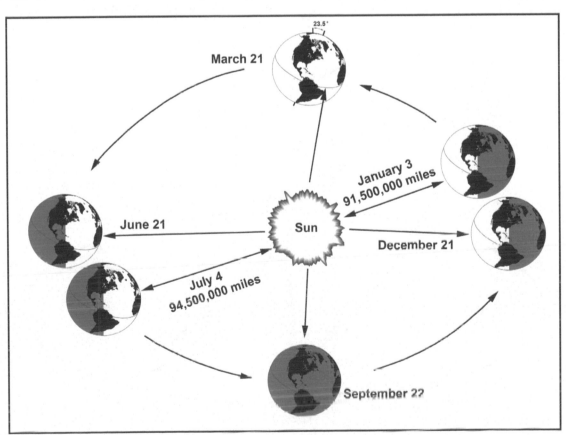

FIGURE 10.1 ▶ *Diagram showing the position of the Earth at various times throughout the year.*

Earth's rotational axis is inclined 23.5° to the plane of Earth's orbit. Earth's rotational axis points to the same location in the sky throughout the year. These two phenomena result in the location on Earth where the sun is exactly overhead (vertical) at noon to vary throughout the year from latitude 23.5° North (**Tropic of Cancer**) to latitude 23.5° South (**Tropic of Capricorn**). The sun is directly overhead the Tropic of Cancer at noon on June 20–21. This is called the **summer solstice** in the northern hemisphere. The sun is directly overhead the Tropic of Capricorn at noon on December 21–22. This is called the **winter solstice** in the northern hemisphere. The sun is directly overhead the equator on March 20–21 (spring or **vernal equinox**) and September 22–23 (fall or **autumnal equinox**). The angle that the sun makes with the Earth directly affects the intensity and duration of sunlight which results in the seasons (winter, spring, summer, and fall).

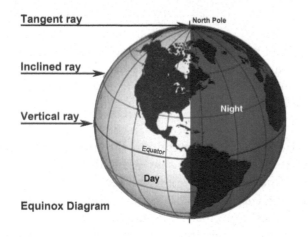

FIGURE 10.2 ▶ *Diagram showing the location of the vertical ray and tangent ray of the sun during the equinoxes. Latitude 66.5° North is called the Arctic Circle. Latitude 66.5° South is called the Antarctic Circle.*

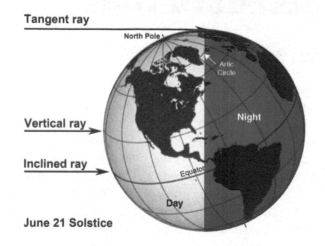

FIGURE 10.3 ▶ *Diagram showing the location of the vertical ray and tangent ray of the sun during the June 21 solstice.*

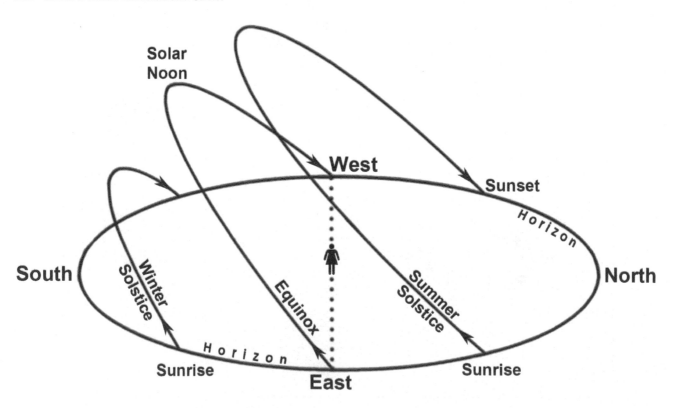

FIGURE 10.4 ▶ *The sun's path through the sky on the winter solstice, equinoxes, and summer solstice. Note that the sun is higher in the sky in summer than it is in winter.*

When the sun is directly overhead (called the **vertical ray**), solar radiation is at its maximum intensity. The farther away from the vertical ray you are, the less intense or more spread out the solar radiation is. At 90° from the vertical ray, the ray is parallel to the surface of the Earth (called the **tangent ray**). From an Earth perspective, when the sun in on the horizon, you are receiving the tangent ray. All other solar rays reach the Earth at an angle between 0° and 90°.

The Sun's position in the sky not only changes throughout the day but also, throughout the year. The number of degrees above the southern horizon the sun would be at solar noon in Ypsilanti, Michigan, on September 23, can be easily calculated. On September 23, the autumnal equinox, the sun's vertical ray is at Latitude 0° (the equator). Ypsilanti is at Latitude 42° 15' North, so Ypsilanti is 42° 15' away from the vertical ray. Therefore, at solar noon, the sun will be 90° − 42° 15' = 47° 45' above the southern horizon.

On the summer solstice, the sun rises 23.5° north of due east and sets north of due west. On the equinoxes, the sun rises due east and sets due west. On the winter solstice, the sun rises 23.5° south of due east and sets south of due west. The sun's path through the sky is longer during the summer than it is during the winter so there are more hours of daylight during summer.

TABLE 10.1 ▶ *Time of sunrise and sunset for Ypsilanti, Michigan on the vernal equinox, summer solstice, autumnal equinox, and the winter solstice for 2011. All times are Daylight Savings Time, except for the winter solstice.*

DATE	SUNRISE	SUNSET
3/20/2011	7:38 am	7:46 pm
6/21/2011	5:48 am	9:14 pm
9/23/2011	7:23 am	7:30 pm
12/21/2011	8:00 am	5:05 pm

EARTH'S SHAPE AND SIZE

Today, no educated person would refute the Earth's general shape or its size since we can travel into outer space and look back on the Earth. However, it does not take advanced technology to accurately determine the Earth's shape and size. The size and shape of the Earth have been well established for over 2,000 years. The ancient Greeks noted that during a lunar eclipse the earth's shadow on the moon is circular. The moon and the sun are spherical so it was easily accepted that the earth was also spherical.

Image © RCPPHOTO, 2011. Modified by Michael D. Bradley. Used under license from Shutterstock.com

FIGURE 10.5 ▶ *A sailing ship approaches over the curve of Earth's surface. Note the top of the ship appears first.*

Ancient people also noted that as a far off ship approaches, the crow's nest appears first, followed by the sails, then by the deck. This observation is easily explained by a spherical earth.

The earth rotates on an axis. Since the earth is spinning, the stars move across the sky in a circular pattern. The center of the circle coincides with the projection of the spin axis of the earth, the North and South Poles, into the sky. We refer to the star that is closest to the projection of Earth's spin axis as the pole star. The brightest star that is closest to the spin axis is Polaris, so we call Polaris the North Star.

If you look at how many degrees of arc that Polaris is above the horizon, you will notice it changes with latitude. At the North Pole, Polaris is directly overhead, and at the equator, Polaris is on the horizon; so in the northern hemisphere, the number of degrees Polaris is above the horizon is approximately, your latitude. This is easily explained by a spherical earth.

The circumference of the Earth was accurately determined by Eratosthenes (276–195 B.C.). The basic idea is this: since the Earth is spherical, there is only one latitude on any given day at which the sun is directly overhead at solar noon. If the sun is directly overhead, no shadow will be cast. As you move north or south of that latitude, the sun will not be directly overhead at noon so a shadow will be cast. The farther you travel, the longer the shadow will be.

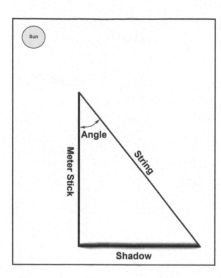

FIGURE 10.6 ▶ *Diagram of Eratosthenes' method of calculating the circumference of the Earth. See text for discussion.*

The circumference of the Earth can be easily calculated by noting the length of the shadows at two different locations along a single meridian. For example, if you held a meter stick vertically on the ground at the equator at noon on the equinox, the sun would be directly overhead, so the stick would cast no shadow. If you took the same measurement at the same time 1000 kilometers north of the equatorial location, the meter stick would cast a shadow 158 millimeters long. If you stretched a string from the top of the meter stick (1000 millimeters long) to the end of the shadow (figure 10.4), you would discover the angle between the meter stick and the string was 9°. There are 360° in a circle so 9° represents 1/40th of a circle. Therefore, the circumference of the Earth

as measured through the North and South Poles is 40 times 1000 kilometers or, 40,000 kilometers.

STANDARD TIME

Solar noon is defined as the time when the sun reaches its highest point in the sky for a given day. The Earth is roughly spherical so solar noon changes with longitude. Solar noon in Detroit is later than solar noon in Cleveland, Ohio and earlier than solar noon in Chicago, Illinois. As transportation became faster, it became necessary to standardize time throughout the world so we could set train schedules without having to deal with thousands of local time differences.

The surface of the Earth is divided into 24 time zones, each approximately 15° of longitude wide. There are local variations in the width of any given time zone to accommodate political considerations. The reference for standard time is the meridian that passes through the Royal Observatory in Greenwich, England. The time in Greenwich, England is referred to as **Universal Time** (formerly called Greenwich Mean Time). Most countries use local time (the 24 time zones) using the midpoint of the time zone as the reference for local time.

Mean noon is based upon standard time so is when the clock reads 12:00 pm*, not when the sun is at its highest point in the sky for a given location, so is different than solar noon. *Technically, as defined by the United States government, noon is the last minute of the morning, 12:00 am, and midnight is the last minute of the evening, 12:00 pm, but common usage is the opposite.

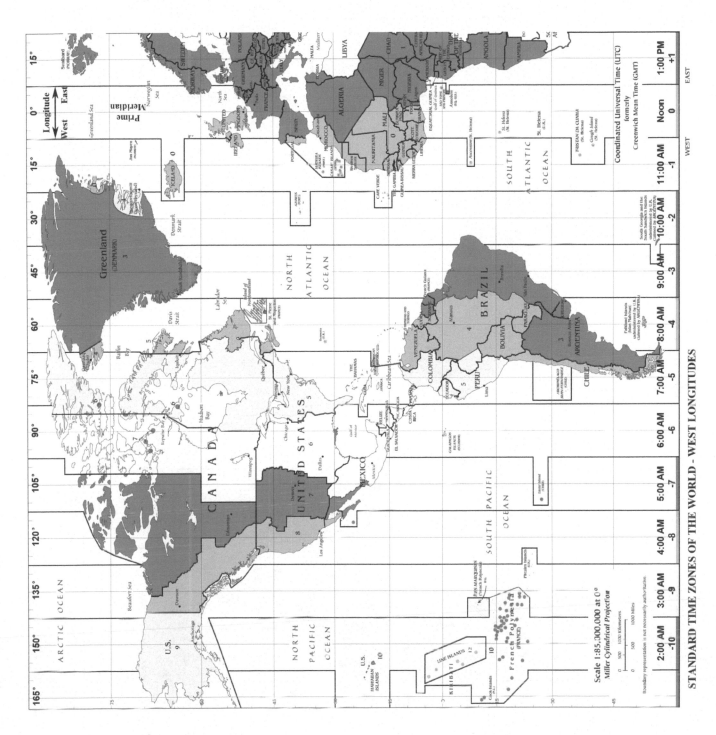

STANDARD TIME ZONES OF THE WORLD - WEST LONGITUDES

FIGURE 10.7 ▲ *Standard Times Zones of the world—Western Hemisphere.*

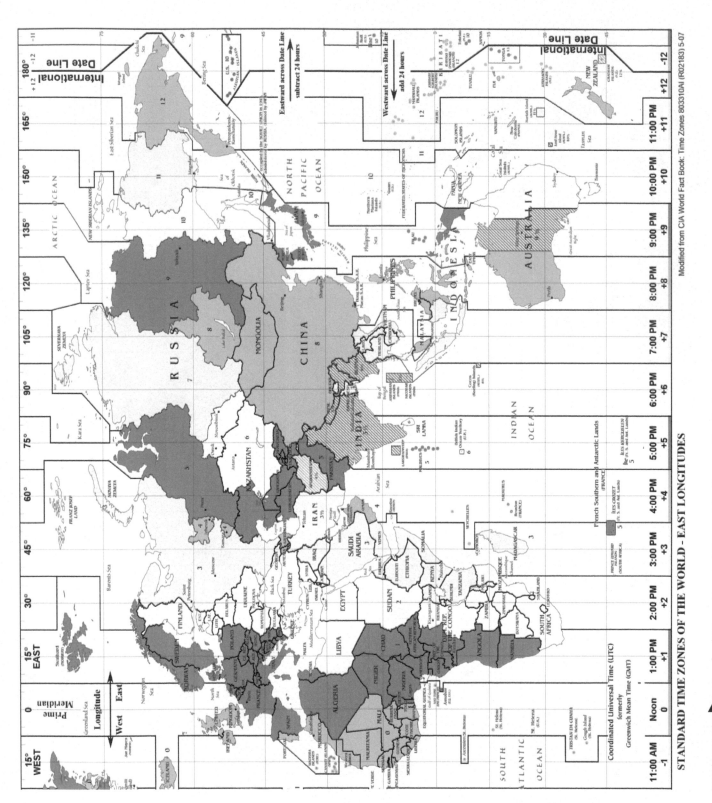

STANDARD TIME ZONES OF THE WORLD - EAST LONGITUDES

FIGURE 10.8 ▲ *Standard Times Zones of the world—Eastern Hemisphere.*

Modified from CIA World Fact Book: Time Zones 803310AI (R02183) 5-07

PreLab
WORKSHEET

NAME _____ DATE _____

Are the seasons a result of the Earth's rotation or revolution?

Is day and night a result of the Earth's rotation or revolution?

Calculate the number of degrees above the southern horizon the sun would be at solar noon in Ypsilanti, Michigan, on the vernal equinox. Show your calculation, including units.

Using Table 10.1, Calculate the hours of daylight for December 21, 2011, the winter solstice, for Ypsilanti, Michigan. Show your calculation, including units.

Chapter 11

Weather

TEMPERATURE

Temperature is a measure of the amount of heat energy in the atmosphere. It varies in a daily cycle in response to heat gain from the sun and heat loss into outer space. When heat gain exceeds heat loss, the temperature rises. When heat loss exceeds heat gain, the temperature falls.

HUMIDITY

Relative humidity is the amount of moisture in the air relative to the maximum amount of moisture the air could hold at a given temperature. The higher the temperature, the more moisture the air can hold. Relative humidity is expressed as a percent. When the amount of moisture in the air is equal to the maximum amount of moisture the air can hold at a given temperature, the relative humidity is 100% (called the **dew point**).

Hygrothermographs are continuous recordings of daily temperature and relative humidity (figure 11.1). The **daily range** is the maximum minus the minimum.

BAROMETRIC PRESSURE

Air pressure (**barometric pressure**) is the weight of atmosphere above a certain point. At sea level the weight of atmosphere is 1013.2 millibars (14.7 lbs/in^2). The instrument used to measure atmospheric pressure is a barometer. The traditional barometer uses mercury. The height of the column of mercury varies with atmospheric pressure so in the United States it has become customary to refer to the atmospheric pressure at any given time as the height (in inches) of mercury in a mercury barometer. Standard atmospheric pressure at sea level (1013.2 millibars) corresponds to 29.92 inches of mercury.

Wind (the horizontal movement of air) is a result of horizontal differences in atmospheric pressure. Air flows from places of high pressure to places with lower pressure. Pressure differences result from unequal heating of the earth's surface. Low pressure centers are called **cyclones** and high pressure centers are called **anticyclones.** **Isobars** (lines of equal air pressure) are drawn on weather maps just like contours (lines of equal elevation) are drawn on topographic maps. The steepness of the slope is indicative of the relative pressure differences between two areas. The greater the pressure differences between two areas, the greater the wind speed.

The wind direction and speed on a weather map is indicated by a flag-like symbol (figure 11.5). The direction the wind is blowing is from the end of the long bar into the small circle. Wind direction is always stated as the direction the wind is coming from. The small circle indicates the location at which the reading is taken. Speed is indicated by flags added to the wind direction pole.

AIR MASSES AND FRONTS

If air stagnates over a region for a period of time it takes on the temperature and moisture characteristics of the region. A large body of air (several kilometers thick and 1600 kilometers or more across) characterized by a homogeneity of temperature and moisture at any given altitude is referred to as an **air mass.**

The boundary between air masses of different densities is called a **front.** As one air mass moves into another some minor mixing occurs along the front but, for the most part, each air mass maintains its identity. Fronts are typically only 15–200 kilometers wide so are represented by broad lines on weather maps. The boundaries are not vertical but slope at a gentle angle with the warmer air mass always over-riding the cooler air mass.

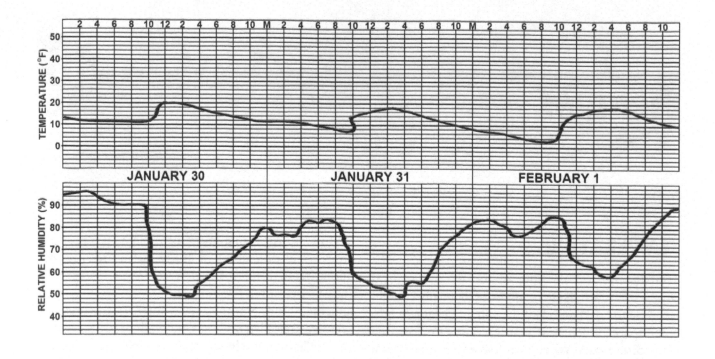

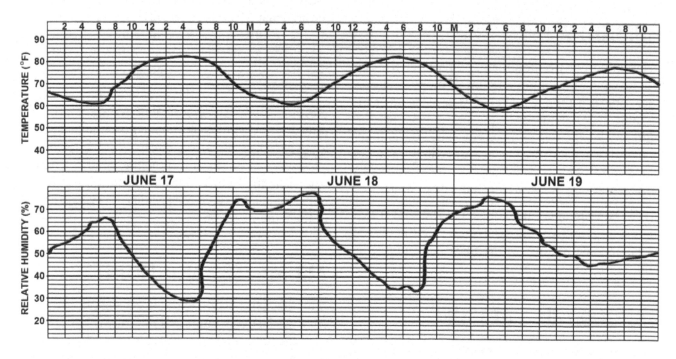

FIGURE 11.1 ▶ *Typical hygrothermographs for summer and winter in a temperate region. M is midnight.*

Temperature decreases with altitude and the ability of our atmosphere to hold moisture is temperature dependent. Therefore, as a warm air mass is forced to climb up the wedge of a cool air mass, the warm air mass cools and clouds form.

If warm air is moving into an area and displacing cooler air, the boundary is called a **warm front**. The average slope of a warm front is 1:200. On a weather map, a warm front is denoted by a line with semicircles extending into the cooler air (figure 11.2).

If cold air is moving into an area and displacing warmer air, the boundary is called a **cold front**. On average, a cold front has a slope of 1:100, so is twice as steep as a warm front. Cold fronts also advance more rapidly than warm fronts. The steeper slope and more rapid advance results in cold fronts having more severe weather associated with them. A cold front produces the same amount of lifting as a warm front but over a shorter distance, so the intensity of precipitation is greater but the duration is shorter. On weather maps, cold fronts are indicated as lines with triangles extending into the warm air mass (figure 11.3).

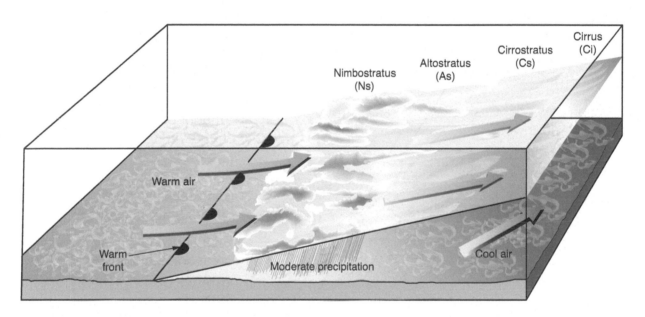

FIGURE 11.2 ▶ *Warm front. See text for discussion.*

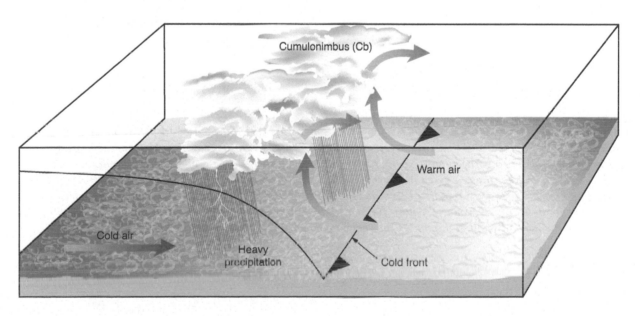

FIGURE 11.3 ▶ *Cold front. See text for discussion.*

Cold fronts generally move faster than warm fronts, so a warm air mass may get squeezed between two cold air masses, which forces the warm air mass to rise until the cold front meets the warm front. This process is called occlusion and the front is called an **occluded front**. On weather maps, occluded fronts are indicated as lines with both semicircles and triangles on the same side.

If the flow of air on both sides of the front is parallel to the front, neither air mass moves into the other, so the front remains stationary. This is called a **stationary front**. On weather maps, stationary fronts are indicated as lines with semicircles on one side and triangles on the other.

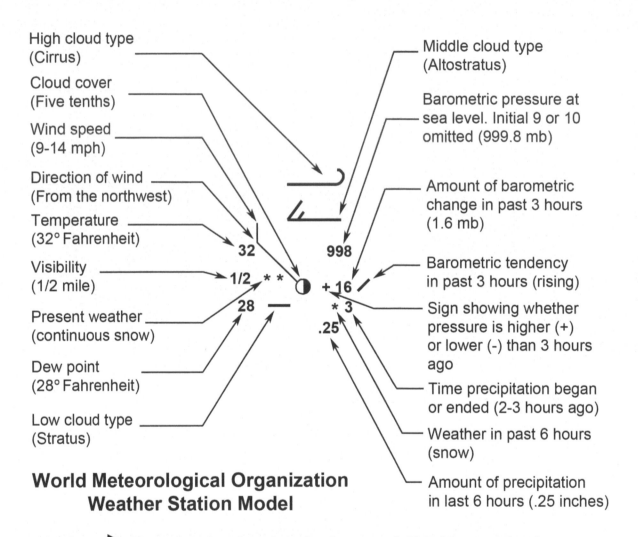

World Meteorological Organization Weather Station Model

FIGURE 11.4 ▶ *Observations of weather at many locations across the United States are plotted on maps providing a basis for spatial evaluations. Map symbols, as well as temperature and dew point information, always appear in the same fixed position on a weather map (called a weather station model). An explanation of the symbols used on weather maps is given in figure 11.5. Note that the barometric pressure is given to tenths of a millbar, and the initial 9 or 10 is omitted. Typical barometric pressures range from 950 to 1050 millibars, so if the first two digits are 50–99, you can assume the omitted first number is 9, and if the first two digits are 00–49, you can assume the omitted first number is 10. Therefore, 998 is 999.8 millibars and 134 is 1013.4 millibars.*

Winds mph		Sky Coverage		Barometric Tendency	
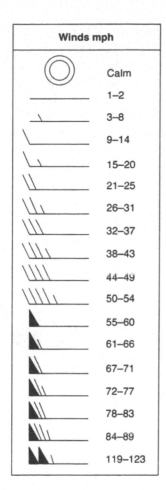	Calm	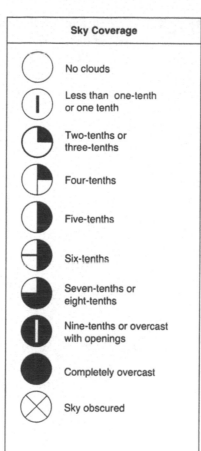	No clouds	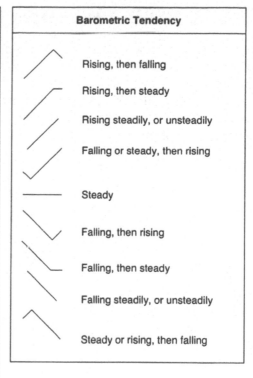	Rising, then falling
	1–2		Less than one-tenth or one tenth		Rising, then steady
	3–8				Rising steadily, or unsteadily
	9–14		Two-tenths or three-tenths		Falling or steady, then rising
	15–20		Four-tenths		
	21–25				Steady
	26–31		Five-tenths		Falling, then rising
	32–37		Six-tenths		Falling, then steady
	38–43		Seven-tenths or eight-tenths		Falling steadily, or unsteadily
	44–49		Nine-tenths or overcast with openings		Steady or rising, then falling
	50–54		Completely overcast		
	55–60		Sky obscured		
	61–66				
	67–71				
	72–77				
	78–83				
	84–89				
	119–123				

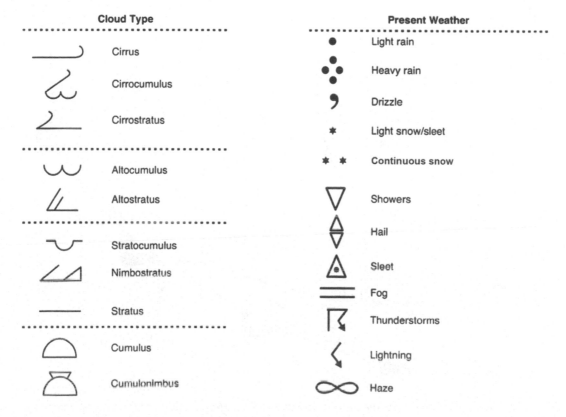

Cloud Type

Cirrus

Cirrocumulus

Cirrostratus

Altocumulus

Altostratus

Stratocumulus

Nimbostratus

Stratus

Cumulus

Cumulonimbus

Present Weather

Light rain

Heavy rain

Drizzle

Light snow/sleet

Continuous snow

Showers

Hail

Sleet

Fog

Thunderstorms

Lightning

Haze

FIGURE 11.5 ▶ *An explanation of symbols used on weather maps.*

HOW TO READ A WEATHER MAP

Figure 11.6 is a surface weather map from June 5, 2005. The heavy lines are isobars labeled with the barometric pressure in millibars. Note there is a high pressure center, an anticyclone, centered over Colorado. There is a cold front along the east and southeast parts of the map.

The current weather at Amarillo, Texas. The current temperature is 52°F and the dew point is 47°F. Visibility is 10 miles. The wind is blowing from the northeast at 3–8 miles per hour. The barometric pressure is 1009.8 millibars. The barometric pressure has changed 0.7 millibars during the last three hours; during this time, the barometric pressure fell or was steady and then rose. Half (5/10) of the sky is covered by clouds.

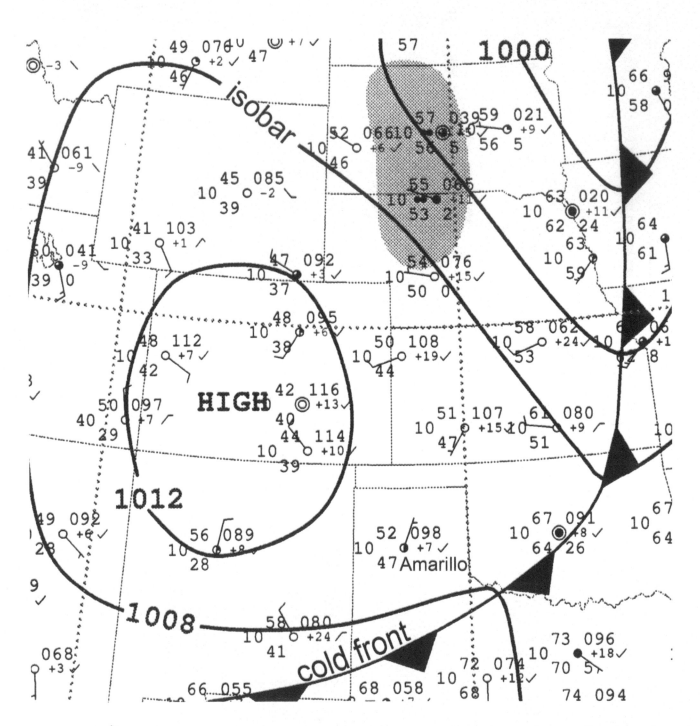

FIGURE 11.6 ▶ *6/5/05 Surface Weather Map and Station Weather at 7:00 AM EST. See text for an explanation on how to read this map.*

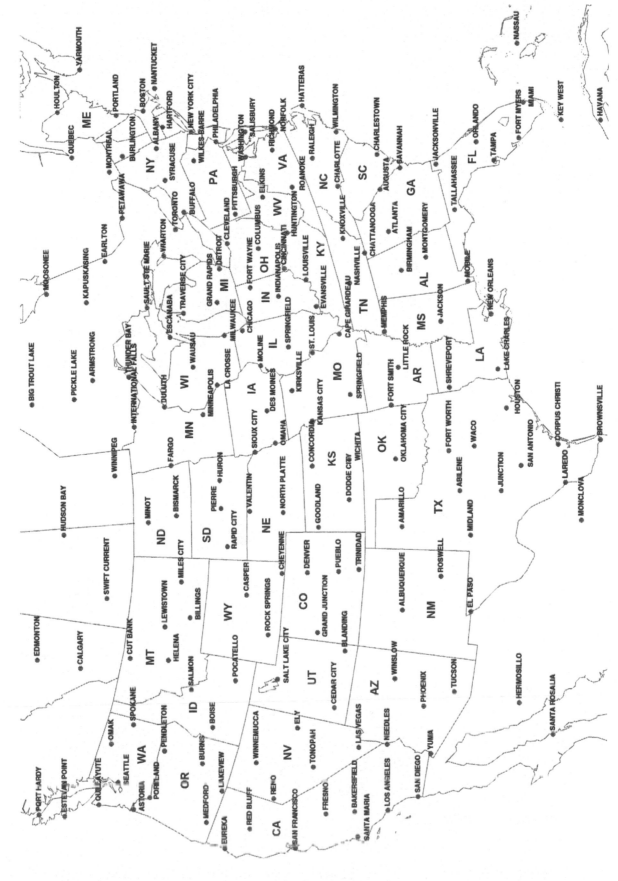

FIGURE 11.7 ▶ *Daily Weather Map Station Names and Locations.*

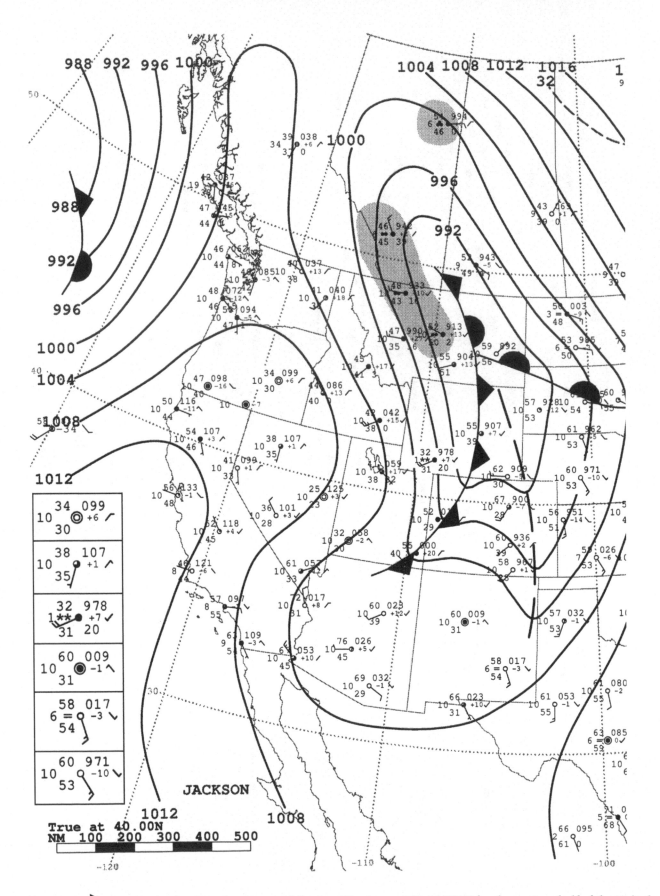

FIGURE 11.8 ▶ *5/17/05 Surface Weather Map and Station Weather at 7:00 AM EST for the western half of the United States. The data for several cities has been enlarged for easier reading and appears in boxes on the left side of the map.*

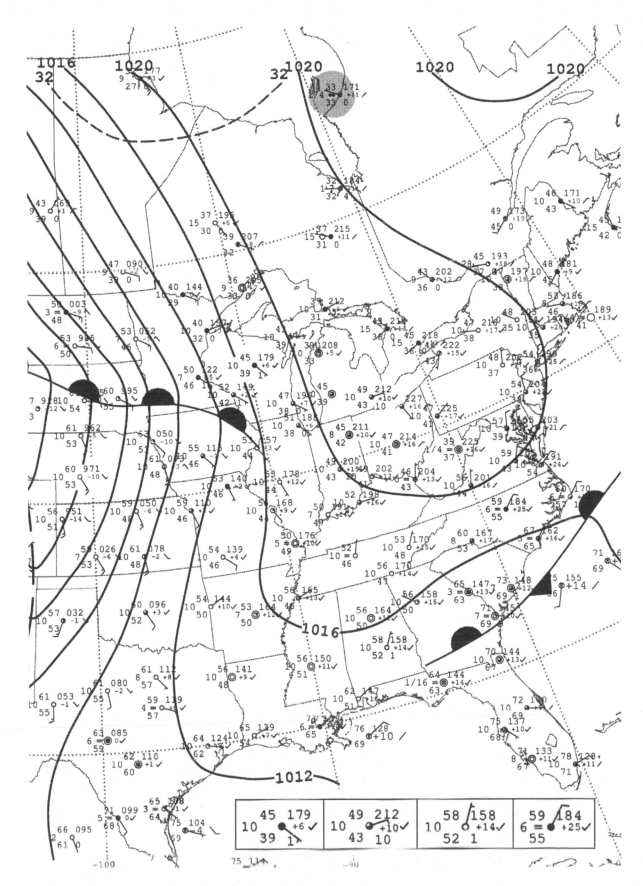

FIGURE 11.9 ▶ *5/17/05 Surface Weather Map and Station Weather at 7:00 AM EST for the eastern half of the United States. The data for several cities has been enlarged for easier reading and appears in boxes at the bottom of the map.*

NAME _____ DATE _____

Using figure 11.1, what is the temperature at 2:00 am on January 30?

Using figure 11.1, what is the relative humidity at 2:00 am on January 30?

Using the 5/17/05 Surface Weather Map and Station Weather at 7:00 AM EST, what is the temperature in Roswell, New Mexico?

Using the 5/17/05 Surface Weather Map and Station Weather at 7:00 AM EST, what is the wind speed in Roswell, New Mexico?

Using the 5/17/05 Surface Weather Map and Station Weather at 7:00 AM EST, what is the dew point in Roswell, New Mexico?

Second Edition

Investigations in **EARTH SCIENCE**

Lab Manual

Michael D. Bradley

Kendall Hunt
publishing company

ISBN 978-1-4652-1402-7

90000

9 781465 214027

43140203

Second Edition

Investigations in **EARTH SCIENCE**

Lab Manual

Michael D. Bradley